TI-83 Plus Graphing Calculator

Calculator

FOR

DUMMIES®

TI-83 Plus Graphing Calculator

FOR DUMMIES®

by C. C. Edwards

WILEY

Wiley Publishing, Inc.

TI-83 Plus Graphing Calculator For Dummies®

Published by
Wiley Publishing, Inc.
111 River Street
Hoboken, NJ 07030-5774

Copyright © 2004 by Wiley Publishing, Inc., Indianapolis, Indiana

Published by Wiley Publishing, Inc., Indianapolis, Indiana

Published simultaneously in Canada

For general information on our other products and services or to obtain technical support, please contact
our Customer Care Department within the U.S. at 800-762-2974, outside the U.S. at 317-572-3993, or fax
317-572-4002.

Wiley also publishes its books in a variety of electronic formats. Some content that appears in print may
not be available in electronic books.

Library of Congress Control Number: 2003114874

ISBN: 0-7645-4970-7

Manufactured in the United States of America

10 9 8 7 6 5 4

1B/QZ/RS/QT/IN

WILEY

About the Author

C. C. Edwards has a Ph.D. in mathematics from the University of Wisconsin, Milwaukee, and is currently teaching mathematics on the undergraduate and graduate levels. She has been using technology in the classroom since before Texas Instruments came out with their first graphing calculator, and she frequently gives workshops at national and international conferences on using technology in the classroom. She has written forty activities for the Texas Instruments Explorations Web site, and she was an editor of *Eightysomething*, a newsletter that used to be published by Texas Instruments. (She still hasn't forgiven TI for canceling that newsletter.)

Just barely five feet tall, CC, as her friends call her, has three goals in life: to be six inches taller, to have naturally curly hair, and to be independently wealthy. As yet, she is nowhere close to meeting any of these goals. When she retires, she plans to become an old lady carpenter.

Dedication

This book is dedicated to Mr. August, my sixth-grade teacher.

Author's Acknowledgments

I'd like to thank Connie Hughes, business development manager at Texas Instruments, for suggesting that I write this book and for helping me get answers to my technical questions. And the folks at John Wiley & Sons who deserve special thanks are Melody Layne, acquisitions editor, and Kala Schrager and Christopher Morris, project editors. Melody and Kala very quickly taught me that a *For Dummies* book is a very special type of book. They were like the math student who always asks, "What's this used for?" Chris explained the ropes to me, kept me on schedule, and gave me extremely good criticism. In fact, I think it is safe to say that these three people have made me a better teacher.

On the home front I'd like to thank Stephen Shauger for volunteering to take over my teaching duties so that I could take the summer off to write this book. And I give many, many thanks to my math soulmates Ioana Mihaila, Olcay Akman, and Fusun Akman for the stimulating conversations and great friendship. I also thank their children Iulia, Cornelia, and Devin for the many happy times we have had together. A special thanks goes to Devin for lending me his TI-83 calculator.

Publisher's Acknowledgments

We're proud of this book; please send us your comments through our online registration form located at www.dummies.com/register/.

Some of the people who helped bring this book to market include the following:

Acquisitions, Editorial, and Media Development

Project Editor: Christopher Morris

Acquisitions Editor: Melody Layne

Senior Copy Editor: Barry Childs-Helton

Technical Editor: Dr. Douglas Shaw, University of Northern Iowa

Editorial Manager: Kevin Kirschner

Permissions Editor: Laura Moss

Media Development Supervisor: Richard Graves

Editorial Assistant: Amanda Foxworth

Cartoons: Rich Tennant, www.the5thwave.com

Production

Project Coordinator: Maridee Ennis

Layout and Graphics: Andrea Dahl, Joyce Haughey, Stephanie D. Jumper, Heather Ryan, Jacque Schneider, Shae Lynn Wilson

Proofreading: TECHBOOKS Production Services

Indexing: TECHBOOKS Production Services

Publishing and Editorial for Technology Dummies

    Richard Swadley, Vice President and Executive Group Publisher

    Andy Cummings, Vice President and Publisher

    Mary C. Corder, Editorial Director

Publishing for Consumer Dummies

    Diane Graves Steele, Vice President and Publisher

    Joyce Pepple, Acquisitions Director

Composition Services

    Gerry Fahey, Vice President of Production Services

    Debbie Stailey, Director of Composition Services

Contents at a Glance

Introduction 1

Part I: Making Friends with the Calculator 7
  Chapter 1: Coping with the Basics 9
  Chapter 2: Doing Basic Arithmetic 21

Part II: Advanced Functions, Complex
Numbers, and Solving Equations 29
  Chapter 3: The Math and Angle Menus 31
  Chapter 4: Dealing with Complex Numbers 41
  Chapter 5: Solving Equations 45

Part III: Dealing with Finances 51
  Chapter 6: Finding the Best Deal 53
  Chapter 7: Loans and Mortgages 57
  Chapter 8: Savings and Investments 63

Part IV: Graphing and Analyzing Functions 67
  Chapter 9: Graphing Functions 69
  Chapter 10: Exploring Functions 83
  Chapter 11: Evaluating Functions 95
  Chapter 12: Drawing on a Graph 105

Part V: Sequences, Parametric Equations,
and Polar Equations 115
  Chapter 13: Graphing Sequences 117
  Chapter 14: Exploring Sequences 127
  Chapter 15: Parametric Equations 139
  Chapter 16: Polar Equations 157

Part VI: Probability and Statistics 177
  Chapter 17: Probability 179
  Chapter 18: Dealing with Statistical Data 183
  Chapter 19: Analyzing Statistical Data 191

Part VII: Dealing with Matrices 203
  Chapter 20: Creating and Editing Matrices 205
  Chapter 21: Using Matrices 211

Part VIII: Communicating with PCs and Other Calculators 217

Chapter 22: Communicating with a PC with TI Connect™ 219
Chapter 23: Communicating Between Calculators 223

Part IX: The Part of Tens 229

Chapter 24: Ten Great Applications 231
Chapter 25: Eight Common Errors 235
Chapter 26: Eleven Common Error Messages 239

Index 243

Table of Contents

Introduction *1*

    About This Book 1

    Conventions Used in This Book 2

    What You're Not to Read 2

    Foolish Assumptions 3

    How This Book Is Organized 3

        Part I: Making Friends with the Calculator 3

        Part II: Advanced Functions, Complex Numbers,

            and Solving Equations 3

        Part III: Dealing with Finances 3

        Part IV: Graphing and Analyzing Functions 4

        Part V: Sequences, Parametric Equations,

            and Polar Equations 4

        Part VI: Probability and Statistics 4

        Part VII: Dealing with Matrices 4

        Part VIII: Communicating with PCs and

            Other Calculators 4

        Part IX: The Part of Tens 4

    Icons Used in This Book 5

    Where to Go from Here 5

Part 1: Making Friends with the Calculator *7*

Chapter 1: Coping with the Basics **9**

    When to Change the Batteries 9

    Turning the Calculator On and Off 10

    Using the Keyboard 10

        Accessing the Functions in Yellow 11

        Using the ALPHA key to write words 11

        Using the ENTER key 12

        Using the X,T,Θ,*n* key 12

        Using the Arrow keys 12

    What Is the Home Screen? 13

    The Busy Indicator 13

    Editing Entries 13

    Using Menus 14

        Accessing a menu 14

        Scrolling a Menu 15

        Selecting Menu Items 15

    Setting the Mode 16

    Using the Catalog 18

Chapter 2: Doing Basic Arithmetic 21

Entering and Evaluating Expressions 21
Important Keys 22
Order of Operations 23
Using the Previous Answer 24
Recycling the Last Entry 25
Storing Variables 26
Combining Expressions 27

Part II: Advanced Functions, Complex Numbers, and Solving Equations *29*

Chapter 3: The Math and Angle Menus 31

The Math Menu and Submenus 31
    Using Math menu functions 32
    Inserting a Math menu function 32
    The Math MATH submenu 33
    The Math NUM submenu 35
The Angle Menu 36
    Converting degrees to radians 37
    Converting radians to degrees 38
    Converting between degrees and DMS 38
    Entering angles in DMS measure 39
    Overriding the mode of the angle 39

Chapter 4: Dealing with Complex Numbers 41

Setting the Mode 41
Using Complex Numbers 42
The Math CPX Submenu 42
    Finding the conjugate of a complex number 43
    Finding the real or imaginary part
       of a complex number 43
    Finding the polar angle and modulus (magnitude)
       of a complex number 43
    Converting between rectangular and polar form 44

Chapter 5: Solving Equations 45

Using the Equation Solver 45
    Step 1. Set the mode 46
    Step 2. Enter or edit the equation to be solved 46
    Step 3. Assign values to variables 47
    Step 4. Define the solution bounds 47
    Step 5. Guess a solution 48
    Step 6. Solve the equation 49
Finding Multiple Solutions 49

Part III: Dealing with Finances *51*

Chapter 6: Finding the Best Deal **53**

Finding the Best Interest Rate 53
Finding the effective rate 53
Finding the nominal rate 54
Leasing versus Borrowing 55

Chapter 7: Loans and Mortgages **57**

Using the TVM Solver 57
Using a TVM Value 60
Finding Principal and Interest 60
Finding the Balance 61

Chapter 8: Savings and Investments **63**

Reaching Financial Goals 63
Finding Future Value of Money 64
Finding Present Value of Money 65

Part IV: Graphing and Analyzing Functions *67*

Chapter 9: Graphing Functions **69**

Entering Functions 69
Graphing Functions 70
Graphing Several Functions 73
Is Your Graph Accurate? 76
Piecewise-Defined Functions 77
Graphing Trig Functions 79
Viewing the Function and Graph on the Same Screen 79
Saving and Recalling a Graph 80

Chapter 10: Exploring Functions **83**

Using Zoom Commands 83
Tracing a Graph 86
Displaying Functions in a Table 88
Clearing a Table 91
Viewing the Table and the Graph on the Same Screen 92

Chapter 11: Evaluating Functions **95**

Finding the Value of a Function 95
Finding the Zeros of a Function 97
Finding Min & Max 98
Finding Points of Intersection 99
Finding the Slope of a Curve 100
Evaluating a Definite Integral 102

Chapter 12: Drawing on a Graph 105

Drawing Lines, Circles, Tangents,
  and Functions on a Graph 106
    Drawing line segments 106
    Drawing horizontal and vertical lines 107
    Drawing circles on a graph 107
    Drawing tangents on a graph 108
    Drawing functions on a graph 108
    Drawing the inverse function 109
  Shading Between Functions 109
  Writing Text on a Graph 111
  Freehand Drawing on a Graph 112
  Erasing Drawings 112
  Saving Graphs and Drawings 113

Part V: Sequences, Parametric Equations, and Polar Equations 115

Chapter 13: Graphing Sequences 117

Entering a Sequence 117
Graphing Sequences 121
Graphing Several Sequences 125
Saving a Sequence Graph 126
Drawing on a Sequence Graph 126

Chapter 14: Exploring Sequences 127

Exploring Sequence Graphs 127
    Using ZOOM in Sequence mode 127
    Tracing a sequence 129
Displaying Sequences in a Table 132
Clearing a User-Defined Table 135
Viewing the Table and the Graph on the Same Screen 136
Evaluating Sequences 137

Chapter 15: Parametric Equations 139

Entering Parametric Equations 139
Graphing Parametric Equations 141
    Graphing several equations 144
    Using ZOOM commands 146
    Saving a parametric graph 148
    Tracing a parametric graph 148
Displaying Equations in a Table 149
    Clearing a user-defined table 152
    Viewing the table and the graph
      on the same screen 153

Evaluating Parametric Equations 154
Finding Derivatives 155

Chapter 16: Polar Equations 157

Converting Coordinates 157
Entering Polar Equations 160
Graphing Polar Equations 161
    Graphing several equations 164
    Using ZOOM commands 165
    Saving a polar graph 167
    Tracing a polar graph 167
Displaying Equations in a Table..169
    Clearing a user-defined table 172
    Viewing the table and the graph
        on the same screen 172
Evaluating Polar Equations 174
Finding Derivatives 175

Part VI: Probability and Statistics 177

Chapter 17: Probability 179

Permutations and Combinations 179
Generating Random Numbers 180
    Generating random integers 180
    Generating random decimals 180

Chapter 18: Dealing with Statistical Data 183

Entering Data 183
Deleting and Editing Data 184
Creating User-Named Data Lists 185
Using Formulas to Enter Data 187
Saving and Recalling Data Lists 188
Sorting Data Lists 189

Chapter 19: Analyzing Statistical Data 191

Plotting One-Variable Data 191
    Constructing a histogram 192
    Constructing a box plot 194
Plotting Two-Variable Data 195
Tracing Statistical Data Plots 195
Analyzing Statistical Data 197
    One-variable data analysis 198
    Two-variable data analysis 199
Regression Models 199

Part VII: Dealing with Matrices 203

Chapter 20: Creating and Editing Matrices 205

Defining a Matrix 205
Editing a Matrix 207
Displaying Matrices 207
Augmenting Two Matrices 207
Copying One Matrix to Another 208
Deleting a Matrix from Memory 209

Chapter 21: Using Matrices 211

Matrix Arithmetic 211
Finding the Determinant 214
Solving a System of Equations 215

Part VIII: Communicating with PCs
and Other Calculators 217

Chapter 22: Communicating with a PC
with TI Connect™ 219

Downloading TI Connect 219
Installing and Running TI Connect 220
Connecting Calculator and PC 220
Transferring Files 220
Upgrading the OS 221

Chapter 23: Communicating Between
Calculators 223

Linking Calculators 223
Transferring Files 224
Transferring Files to Several Calculators 226

Part IX: The Part of Tens 229

Chapter 24: Ten Great Applications 231

Ten Great Applications 232
Downloading an Application 233
Installing an Application 233

Chapter 25: Eight Common Errors 235

Using $-$ Instead of $(\cdot)$ to Indicate That
   a Number Is Negative 235
Indicating the Order of Operations Incorrectly
   by Using Parentheses 235

Improperly Entering the Argument
for Menu Functions 236
Entering an Angle in Degrees While in Radian Mode 236
Graphing Trigonometric Functions
While in Degree Mode 236
Graphing Functions When Stat Plots Are Active 237
Graphing Stat Plots When Functions
or Other Stat Plots Are Active 237
Setting the Window Inappropriately for Graphing 237

Chapter 26: Eleven Common Error Messages 239

ARGUMENT 239
BAD GUESS 239
DATA TYPE 240
DIM MISMATCH 240
DOMAIN 240
INVALID 240
INVALID DIM 240
NO SIGN CHNG 241
SINGULAR MAT 241
SYNTAX 241
WINDOW RANGE 241

Index *243*

Introduction

● ●

Do you know how to use the TI-83, TI-83 Plus, TI-83 Plus Silver Edition, TI-84 Plus, or TI-84 Plus Silver Edition graphing calculator to do each of the following?

✔ Solve equations and systems of equations

✔ Analyze loan options

✔ Graph functions, parametric equations, polar equations, and sequences

✔ Create stat plots and analyze statistical data

✔ Multiply matrices

✔ Write a program

✔ Transfer files between two or more calculators

✔ Save calculator files on your computer

✔ Add applications to your calculator so it can do even more than it could when you bought it

If not, then this is the book for you. Contained within these pages are straightforward, easy-to-follow directions that tell how to do everything listed here — and much, much more.

About This Book

Although this book does not tell you how to do *everything* the calculator is capable of doing, it gets pretty close. It covers more than just the basics of using the calculator, paying special attention to warning you of the problems you could encounter if you knew only the basics of using the calculator.

This is a reference book. It's process-driven, not application-driven. You won't be given a problem to solve and then be told how to use the calculator to solve that particular problem. Instead, you're given the steps needed to get the calculator to perform a particular task, such as constructing a histogram.

Conventions Used in This Book

When I refer to "the calculator," I am referring to the TI-83 Plus and the TI-83 Plus Silver Edition, as well as the TI-83. Sometimes — not often — the TI-83 works differently from the other two Plus calculators. When that's the case, I offer directions for the TI-83 in parentheses.

When I want you to press a key on the calculator, I use an icon for that key. For example, if I want you to press the ENTER key, I say press ENTER. If I want you to press a series of keys, such as the Stat key and then the Right Arrow key, I say (for example) press STAT ▶. All keys on the calculator are pressed one at a time. On the calculator, there is no such thing as holding down one key while you press another key.

It's tricky enough to get handy with the location of the keys on the calculator, and even more of a challenge to remember the location of the secondary functions (the yellow functions above the key). So when I want you to access one of those functions, I give you the actual keystrokes. For example, if I want you to access the Draw menu, I tell you to press 2nd PRGM. This is a simpler method than that of the manual that came with your calculator — which would say press 2nd[DRAW] and then make you hunt for the location of the secondary function DRAW. The same principle holds for using key combinations to enter specific characters; for example, I tell you to press ALPHA 0 to enter a space.

When I want you to use the Arrow keys, but not in any specific order, I say press ▶ ◀ ▲ ▼. If I want you to use only the Up- and Down-Arrow keys, I say press ▲ ▼.

What You're Not to Read

Of course, you don't have to read anything you don't want to. The only items in this book that really don't need to be read are the items that follow a Technical Stuff icon. These items are designed for the curious reader who wants to know, but doesn't really need to know, why something happens.

Other items that you may not need to read are the paragraphs that follow the steps in a procedure. These paragraphs are designed to give you extra help should you need it. The steps themselves are in **bold**; the explanatory paragraphs are in a normal font.

Foolish Assumptions

My nonfoolish assumption is that you know (in effect) nothing about using the calculator, or you wouldn't be reading this book. My foolish assumptions are as follows:

- ✔ You own, or have access to, one of the calculators listed at the beginning of the Introduction.

- ✔ If you want to transfer files from your calculator to your computer, I assume that you have a computer and know the basics of how to operate it.

How This Book Is Organized

The parts of this book are organized by tasks that you would like to have the calculator perform.

Part I: Making Friends with the Calculator

This part describes the basics of using the calculator. It addresses such tasks as adjusting the contrast and getting the calculator to perform basic arithmetic operations.

Part II: Advanced Functions, Complex Numbers, and Solving Equations

Here things get more complicated. This part tells you how to use the many great functions housed in the Math menu, such as the function that converts a decimal to a fraction. This part also tells you how to deal with complex numbers and use the calculator to solve an equation.

Part III: Dealing with Finances

This part tells you how to use the really great Finance application housed in your calculator to do things like calculate the best interest rate and find internal rates of return.

Part IV: Graphing and Analyzing Functions

In this part, think visual. Part IV tells you how to graph and analyze functions, draw on your graph, and create a table for the graph.

Part V: Sequences, Parametric Equations, and Polar Equations

This part describes how you can graph and analyze parametric equations, polar equations, and sequences.

Part VI: Probability and Statistics

It's highly probable that Part VI will tell you how to deal with probability and statistics.

Part VII: Dealing with Matrices

Red pill or blue pill? Part VII takes you deep inside the world of matrices.

Part VIII: Communicating with PCs and Other Calculators

Your calculator joins the information superhighway. Part VIII describes how you can save calculator files on a computer and how you can transfer files from one calculator to another.

Part IX: The Part of Tens

Part IX contains a plethora of wonderful information. This part tells you about the many wonderful applications you can put on your calculator and it describes the most common errors and error messages that you may encounter.

Icons Used in This Book

This book uses three icons to help you along the way. Here's what they are and what they mean:

The text following this icon tells you about shortcut and other ways of enhancing your use of the calculator.

The text following this icon tells you something you should remember because if you don't it may cause you problems later. Usually it's a reminder to enter the appropriate type of number so you can avoid an error message.

There is no such thing as crashing the calculator. But this icon warns you of those *few* times when you can do something wrong on the calculator and be totally baffled because the calculator is giving you confusing feedback — either no error message or a cryptic error message that doesn't really tell you the true location of the problem.

This is the stuff you don't really need to read unless you're really curious.

Where to Go from Here

This book is designed so that you do not have to read it from cover to cover. You don't even have to start reading at the beginning of a chapter. When you want to know how to get the calculator to do something, just start reading at the beginning of the appropriate section. The Index and Table Of Contents should help you find whatever you're looking for.

Part I
Making Friends with the Calculator

The 5th Wave   By Rich Tennant

"IT SAYS HERE IF I SUBSCRIBE TO THIS MAGAZINE, THEY'LL SEND ME A FREE DESK-TOP CALCULATOR. DESKTOP CALCULATOR?!! WHOOAA-WHERE HAVE I BEEN?!!"

In this part...

*T*his part takes you once around the block with the basics of using the calculator. In addition to showing you how to use the calculator to evaluate arithmetic expressions, I discuss the elementary calculator functions — including multi-use keys, menus, modes, and the Catalog. I also cover expressions and the order of operations, storing and recalling variables, and combining expressions.

Chapter 1

Coping with the Basics

- -

In This Chapter

▶ Turning the calculator on and off

▶ Using the keyboard

▶ Using the menus

▶ Setting the mode of the calculator

▶ Using the Catalog

- -

The TI-83 and TI-83 Plus graphics calculators are loaded with many useful features. With them, you can graph and investigate functions, parametric equations, polar equations, and sequences. You can use them to analyze statistical data and to manipulate matrices. You can even use them to calculate mortgage payments.

But if you've never used a graphics calculator before, you may at first find it a bit intimidating. After all, it contains about two dozen menus, many of which contain three or four submenus. But it's really not that hard to get used to using the calculator. After you get familiar with what the calculator is capable of doing, finding the menu that houses the command you need is quite easy. And you have this book to help you along the way.

When to Change the Batteries

The convenience of battery power has a traditional downside: What if the batteries run out of juice at a crucial moment, say during a final exam? Fortunately, the calculator gives you some leeway. When your batteries are low, the calculator displays a "Your batteries are low" warning message. After you see this message for the first time, the calculator should, according to the manufacturer, continue to function just fine for at least one week. There is one exception: If you attempt to download an application when the batteries are low, the calculator displays a "Batteries are low —

Change is required" warning message and refuses to download the application until after you've changed the batteries. (Chapter 27 explains how to download applications.)

Because you've likely put batteries into countless toys, you should have no trouble opening the cover on the back of the calculator and popping in four AAA batteries. Above the AAA battery chamber is a panel that opens to the compartment containing the backup battery. The type of battery housed in this compartment is indicated on the lid of the panel. The manufacturer recommends that you replace this battery every three or four years. So mark your calendar!

Turning the Calculator On and Off

Press ⌜ON⌝ to turn the calculator on. To turn the calculator off, press ⌜2nd⌝, and then press ⌜ON⌝. These keys are in the left column of the keyboard. The ⌜ON⌝ key is at the bottom of the column, and the ⌜2nd⌝ key is the second key from the top of this column.

To prolong the life of the batteries, the calculator automatically turns itself off after five minutes of inactivity. But don't worry — when you press ⌜ON⌝, all your work will appear on the calculator just as you left it before the calculator turned itself off.

In some types of light, the screen can be hard to see. To increase the contrast, repeatedly press ⌜2nd⌝⌜▲⌝. Because the keys on the calculator must be pressed one at a time, press ⌜2nd⌝ and then press ⌜▲⌝. Continue pressing this combination of keystrokes until you have the desired contrast.

To decrease the contrast, repeatedly press ⌜2nd⌝⌜▼⌝.

Using the Keyboard

The row of keys under the calculator screen contains the keys you use when graphing. The next three rows, for the most part, contain editing keys, menu keys, and arrow keys. The arrow keys (⌜▶⌝⌜◀⌝⌜▲⌝⌜▼⌝) control the movement of the cursor. The remaining rows contain, among other things, the keys you typically find on a scientific calculator.

Keys on the calculator are always pressed one at a time; they are *never* pressed simultaneously. In this book, an instruction such as ⌜2nd⌝⌜ON⌝ indicates that you should first press ⌜2nd⌝ and then press ⌜ON⌝.

Accessing the functions in yellow

Above and to the left of most keys is a secondary key function written in yellow. To access that function, first press 2nd and then press the key. For example, π is in yellow above the ^ key, so to use π in an expression, press 2nd and then press ^.

Because hunting for the function in yellow can be tedious, in this book I use only the actual keystrokes. For example, I will make statements like, "π is entered into the calculator by pressing 2nd ^." Most other books would state, "π is entered into the calculator by pressing 2nd [π]."

When the 2nd key is active and the calculator is waiting for you to press the next key, the blinking ■ cursor symbol is replaced with the ↑ symbol.

Using the ALPHA key to write words

Above and to the right of most keys is a letter written in green. To access these letters, first press ALPHA and then press the key. For example, because the letter O is in green above the 7 key, to enter this letter, press ALPHA and then press 7.

Because hunting for letters on the calculator can be tedious, I tell you the exact keystrokes needed to create them. For example, if I want you to enter the letter O, I say, "Press ALPHA 7 to enter the letter O." Most other books would say "Press ALPHA [O]" and leave it up to you to figure out where that letter is on the calculator.

You must press ALPHA before entering each letter. However, if you want to enter many letters, first press 2nd ALPHA to lock the calculator in Alpha mode. Then all you have to do is press the keys for the various letters. When you're finished, press ALPHA to take the calculator out of Alpha mode. For example, to enter the word TEST into the calculator, press 2nd ALPHA 4 SIN LN 4 and then press ALPHA to tell the calculator that you're no longer entering letters.

When the calculator is in Alpha mode, the blinking ■ cursor symbol is replaced with the ↑ symbol. This symbol indicates that the next key you press will insert the green letter above that key. To take the calculator out of Alpha mode, press ALPHA.

Using the [ENTER] key

The [ENTER] key is used to evaluate expressions and to execute commands. After you have, for example, entered an arithmetic expression (such as 5 + 4), press [ENTER] to evaluate that expression. In this context, the [ENTER] key functions as the equal sign. Entering arithmetic expressions is explained in Chapter 2.

Using the [X,T,Θ,n] key

[X,T,Θ,n] is the key you use to enter the variable in the definition of a function, a parametric equation, a polar equation, or a sequence. In Function mode, this key produces the variable **X**. In Parametric mode it produces the variable **T**; and in Polar and Sequence modes it produces the variables θ and *n*, respectively. Setting the mode is explained later in this chapter.

Using the Arrow keys

The Arrow keys ([▶], [◀], [▲], and [▼]) control the movement of the cursor. These keys are in a circular pattern in the upper-right corner of the keyboard. As expected, [▶] moves the cursor to the right, [◀] moves it to the left, and so on. When I want you to use the Arrow keys — but not in any specific order — I refer to them all together, as in: "Use [▶][◀][▲][▼] to place the cursor on the entry."

Keys to remember

The following keystrokes are invaluable:

✔ [2nd][MODE]: This is the equivalent of the Escape key on a computer. It gets you out of whatever you're doing (or have finished doing) and returns you to the Home screen. The Home screen is where the action takes place. This is where you execute commands and evaluate expressions.

✔ [ENTER]: This key is used to execute commands and to evaluate expressions. When evaluating expressions, it's the equivalent of the equal sign.

✔ [CLEAR]: This is the "erase" key. If you're entering something into the calculator and change your mind, press this key. If you want to erase the contents of the Home screen, repeatedly press this key until the Home screen is blank.

What Is the Home Screen?

The Home screen is the screen that appears on the calculator when you first turn it on. This is the screen where most of the action takes place as you use the calculator — it's where you evaluate expressions and execute commands. This is also the screen you usually return to after you've completed a task such as entering a matrix in the Matrix editor or entering data in the Stat List editor.

Press 2nd MODE to return to the Home screen from any other screen. This combination of keystrokes, 2nd MODE, is the equivalent of the "escape" key on a computer. It always takes you back to the Home screen.

If you want to clear the contents of the Home screen, repeatedly press CLEAR until the Home screen is blank.

The Busy Indicator

If you see a moving vertical line in the upper-right corner of the screen, this indicates that the calculator is busy graphing a function, evaluating an expression, or executing a command.

If it's taking too long for the calculator to graph a function, evaluate an expression, or execute a command, and you want to abort the process, press ON. If you're then confronted with a menu that asks you to select either **Quit** or **Goto,** select **Quit** to abort the process.

Editing Entries

The calculator offers four ways to edit an entry:

  ✔ **Deleting the entire entry:**

   Use ▶◀▲▼ to place the cursor anywhere in the entry and then press CLEAR and to delete the entry.

  ✔ **Erasing part of an entry:**

   To erase a single character, use ▶◀▲▼ to place the cursor on the character you want to delete and then press DEL to delete that character.

✔ **Inserting characters:**

Because "typing over" is the default mode, to insert characters you must first press [2nd][DEL] to enter Insert mode. When you insert characters, the inserted characters are placed to the left of the cursor. For example, if you want to insert CD between B and E in the word ABEF, you would place the cursor on E to make the insertion.

To insert characters, use [▶][◀][▲][▼] to place the cursor at the location of the desired insertion, press [2nd][DEL], and then key in the characters you want to insert. When you're finished inserting characters, press one of the Arrow keys to take the calculator out of Insert mode.

✔ **Keying over existing characters:**

"Type over" is the default mode of the calculator. So if you want to overtype existing characters, just use [▶][◀][▲][▼] to put the cursor where you want to start, and then use the keyboard to enter new characters.

On the Home screen, the calculator doesn't allow you to directly edit an already-evaluated expression or an already-executed command. But you can recall that expression or command if it was the last entry you made in the calculator — and when it's recalled, you can edit it. To recall the last expression or command, press [2nd][ENTER]. This makes the calculator paste a copy of the desired expression or command on the Home screen so you can edit it.

Using Menus

Most functions and commands you use are found in the menus housed in the calculator — and just about every chapter in this book refers to them. This section is designed to give you an overview of how to find and select menu items.

Accessing a menu

Each menu has its own key or key combination. For example, to access the Math menu, press [MATH]; to access the Test menu, press [2nd][MATH]. An example of a menu appears in the first picture in Figure 1-1. This is a picture of the Math menu.

Some menus, such as the Math menu, contain submenus. This is also illustrated in the first picture in Figure 1-1. This picture shows that the submenus in the Math menu are **MATH, NUM, CPX,** and **PRB** (Math, Number, Complex, and Probability). Repeatedly press ▶ to view the items on the other submenus; repeatedly press ◀ to return to the Math MATH submenu. This is illustrated in the second and third pictures in Figure 1-1.

Math MATH menu   Math NUM menu   Math PRB menu

Figure 1-1: Submenus of the Math menu.

Scrolling a menu

After the number 7 in the first two pictures in Figure 1-1, a down arrow indicates that more items are available in the menu than appear on-screen. There's no down arrow after the 7 in the third picture in Figure 1-1 because that menu has exactly seven items.

To see menu items that don't appear on-screen, repeatedly press ▼. To get quickly to the bottom of a menu from the top of the menu, press ▲. Similarly, to quickly get from the bottom to the top, press ▼.

Selecting menu items

To select a menu item from a menu, key in the number (or letter) of the item or use ▼ to highlight the number (or letter) of the item and then press ENTER.

Some menus, such as the Mode menu that is pictured in Figure 1-2, require that you select an item from a list of items by highlighting that item. The list of items usually appear in a single row and the calculator requires that one item in each row be highlighted. To highlight an item, use ▶◀▲▼ to place the cursor on the item and then press ENTER to highlight the item. The selections on the Mode menu are described in the next section.

Setting the Mode

The Mode menu, which is accessed by pressing MODE, is the most important menu on the calculator; it tells the calculator how you want numbers and graphs to be displayed. The Mode menu is pictured in Figure 1-2.

Figure 1-2: The Mode menu.

One item in each row of this menu must be selected. Here are your choices:

✔ **Normal, Sci, or Eng:**

This setting controls how numbers are displayed on the calculator. In Normal mode, the calculator displays numbers in the usual numeric fashion that you used in elementary school — provided it can display it using no more than ten digits. If the number requires more than ten digits, the calculator displays it using scientific notation.

In Scientific (**Sci**) mode, numbers are displayed using scientific notation; and in Engineering (**Eng**) mode, numbers are displayed in engineering notation. These three modes are illustrated in Figure 1-3. In this figure, the first answer is displayed in normal notation, the second in scientific notation, and the third in engineering notation.

In scientific and engineering notation, the calculator uses En to denote multiplication by 10n.

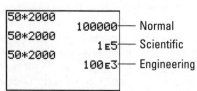

Figure 1-3: Normal, scientific, and engineering notations.

↙ **Float 0123456789:**

Select **Float** if you want the calculator to display as many digits as possible. Select **0** if you want all numbers rounded to an integer. If you're dealing with money, select **2** so that all numbers will be rounded to two decimal places. Selecting **5** rounds all numbers to five decimal places, and, well, you get the idea.

↙ **Radian or Degree:**

If you select **Radian,** all angles entered in the calculator are interpreted as being in radian measure; all angular answers given by the calculator will also be in radian measure. Similarly, if you select **Degree,** any angle you enter must be in degree measure, and any angular answer given by the calculator is also in degree measure.

↙ **Func, Par, Pol, or Seq:**

This setting tells the calculator what type of functions you plan to graph. Select **Func** to graph plain old vanilla functions, $y = f(x)$. Select **Par** to graph parametric equations; **Pol** to graph polar equations; and **Seq** to graph sequences. (Sequences are also called *iterative equations.*)

↙ **Connected or Dot:**

In **Dot** mode, the calculator produces a graph by plotting only the points it calculates. In **Connected** mode, the calculator joins consecutively plotted points with a line.

My recommendation is to select the **Connected** mode because each of the graphing options (**Func, Par, Pol,** and **Seq**) allows you to select a graphing style, one of which is the dot style.

↙ **Sequential or Simul:**

In **Sequential** mode, the calculator completes the graph of one function before it graphs the next function. In Simultaneous **(Simul)** mode, the calculator graphs all functions at the same time. It does so by plotting the values of all functions for one value of the independent variable, and then plotting the values of all functions for the next value of the independent variable.

Simul mode is useful if you want to see whether two functions intersect at the same value of the independent variable. You have to watch the functions as they are graphed in order to *see* if this happens.

✔ **Real, a + b***i***, or re^θ***i*:

If you're dealing with only real numbers, select the **Real** mode. If you're dealing with complex numbers, select **a + b***i* if you want the complex numbers displayed in rectangular form. If you want complex numbers displayed in polar form, select the **re^θ***i* mode.

✔ **Full, Horiz, or G-T:**

The **Full** screen mode displays the screen as you see it when you turn the calculator on. The other screen modes are split-screen modes. The **Horiz** mode is for when you want to display a graph and the Y= editor or the Home screen at the same time. Use the **G-T** mode when you want to display a graph and a table at the same time. (The split-screen modes are explained in detail in Chapters 9, 10, 14, 15, and 16.)

If you're planning on graphing trigonometric functions, put the calculator in Radian mode. Reason: Most trig functions are graphed for $-2\pi \leq x \leq 2\pi$. That is approximately $-6.28 \leq x \leq 6.28$. That's not a bad value for the limits on the *x*-axis. But if you graph in Degree mode, you would need $-360 \leq x \leq 360$ for the limits on the *x*-axis. This is doable . . . but trust me, it's easier to graph in Radian mode.

If your calculator is in Radian mode and you want to enter an angle in degrees, Chapter 3 tells you how to do so without resetting the mode.

Using the Catalog

The calculator's Catalog houses every command and function used by the calculator. However, it's easier to use the keyboard and the menus to access these commands and functions than it is to use the Catalog. There are several exceptions; for example, the hyperbolic functions are found only in the Catalog. If you have to use the Catalog, here's how to do it:

1. **If necessary, use** ▶ ◀ ▲ ▼ **to place the cursor at the location where you want to insert a command or function found in the Catalog.**

   The command or function is usually inserted on the Home screen, or in the Y= editor when you're defining a function you plan to graph.

2. Press 2nd 0 **to enter the Catalog.**

This is illustrated in the first picture in Figure 1-4.

3. Enter the first letter in the name of the command or function.

Notice that the calculator is already in Alpha Mode, as is indicated by the ▣ in the upper-right part of the screen. To enter the letter, all you have to do is press the key corresponding to that letter. For example, if you're using the Catalog to access the hyperbolic function tanh, press 4 because the letter **T** is written in green above this key. This is illustrated in the second picture in Figure 1-4.

4. Repeatedly press ▾ to move the indicator to the desired command or function.

5. Press ENTER **to select the command or function.**

This is illustrated in the third picture in Figure 1-4. After pressing ENTER, the command or function is inserted at the cursor location.

Press 2nd 0   Enter first letter   Select item

Figure 1-4: Steps for using the Catalog.

Chapter 2

Doing Basic Arithmetic

· ·

In This Chapter

▶ Entering and evaluating arithmetic expressions

▶ Obeying the order of operations

▶ Storing and recalling variables

▶ Combining expressions

▶ Writing small programs

· ·

*W*hen you use the calculator to evaluate an arithmetic expression such as $5^{10} + 4^6$, the format in which the calculator displays the answer depends on how you have set the Mode of the calculator. Do you want answers displayed in scientific notation? Do you want all numbers rounded to two decimal places?

Setting the *mode* of the calculator affords you the opportunity to tell the calculator how you want these, and other questions, answered. (Setting the mode is explained in Chapter 1.) When you're doing basic, real-number arithmetic, the mode is usually set as it appears in Figure 1-2 of Chapter 1.

Entering and Evaluating Expressions

Arithmetic expressions are evaluated on the Home screen. The Home screen is the screen you see when you turn the calculator on. If the Home screen is not already displayed on the calculator, press [2nd][MODE] to display it. If you want to clear the contents of the Home screen, repeatedly press [CLEAR] until the screen is empty.

Arithmetic expressions are entered in the calculator the same way you would write them on paper if you were restricted to using the division sign (/) for fractional notation. This restriction sometimes requires parentheses around the numerator or the denominator, as illustrated in the first two calculations in Figure 2-1.

There is a major difference between the subtraction $\boxed{-}$ key and the $\boxed{(-)}$ negation key. They are not the same, nor are they interchangeable. Use the $\boxed{-}$ key to indicate subtraction; use the $\boxed{(-)}$ key before a number to identify that number as negative. If you improperly use $\boxed{(-)}$ to indicate a subtraction problem, or if you improperly use $\boxed{-}$ to indicate that a number is negative, you get the ERR: SYNTAX error message. The use of these two symbols is illustrated in the last calculation in Figure 2-1.

When entering numbers, do not use commas. For example, the number 1,000,000 is entered in the calculator as 1000000.

After entering the expression, press $\boxed{\text{ENTER}}$ to evaluate it. The calculator displays the answer on the right side of the next line, as shown in Figure 2-1.

```
5+3/2
               6.5
(5+3)/2
                 4
-3--4
                 1
```

Figure 2-1: Evaluating arithmetic expressions.

Important Keys

Starting with the fifth row of the calculator, you find the functions commonly used on a scientific calculator. Here's what they are and how you use them:

✔ **π and *e*:**

   The transcendental numbers π and *e* are respectively located in the fifth and sixth rows of the last column of the keyboard. To enter π in the calculator, press $\boxed{\text{2nd}}\boxed{\wedge}$; to enter *e*, press $\boxed{\text{2nd}}\boxed{\div}$, as shown in the first line of Figure 2-2.

✔ **The trigonometric and inverse trigonometric functions:**

   The trigonometric and inverse trigonometric functions are located in the fifth row of the keyboard. These functions require that the argument of the function be enclosed in parentheses. To remind you of this, the calculator provides the first parenthesis for you (as in the first line of Figure 2-2).

✔ **The square-root and exponential functions:**

   The square-root function, 10^x function, and e^x function are respectively located in the sixth, seventh, and eighth rows of the first column on the keyboard. Each of these functions requires that its argument be enclosed in parentheses. To remind you of this, the calculator provides the first parenthesis for you (as in the third line of Figure 2-2).

```
sin⁻¹(cos(π))
          86.85840735
√(3²+4²)
                    5
-3²
                   -9
```

Figure 2-2: Examples of arithmetic expressions.

✔ **The inverse and square functions:**

The inverse and square functions are respectively located in the fifth and sixth rows of the left column on the calculator. To enter the multiplicative inverse of a number, enter the number and then press x^{-1}. Similarly, to square a number, enter the number and then press x^2. The third line of Figure 2-2 shows this operation.

If you want to evaluate an arithmetic expression and you need a function other than those just listed, you'll most likely find that function in the Math menu (described in detail in the next chapter). The hyperbolic functions are an exception; those you find in the calculator's Catalog. (Chapter 1 discusses the Catalog and how to access the hyperbolic functions).

Order of Operations

The order in which the calculator performs operations is the standard order that we are all used to. Spelled out in detail, here is the order in which the calculator performs operations:

1. **The calculator simplifies all expressions surrounded by parentheses.**

2. **The calculator evaluates all functions that are followed by the argument.**

   These functions supply the first parenthesis in the pair of parentheses that must surround the argument. An example is sin x. When you press SIN to access this function, the calculator inserts sin(on-screen. You then enter the argument and press ⌐).

3. **The calculator evaluates all functions entered after the argument.**

   An example of such a function is the square function. You enter the argument and then press $\boxed{x^2}$ to square it.

   Evaluating -3^2 may not give you the expected answer. We think of -3 as being a single, negative number. So when we square it, we expect to get $+9$. But the calculator gets -9 (as indicated in the fifth line of Figure 2-2). This happens because the normal way to enter -3 into the calculator is by pressing $\boxed{(\text{-})}\boxed{3}$ — and pressing the $\boxed{(\text{-})}$ key is equivalent to multiplying by -1. Thus, in this context, $-3^2 = -1 * 3^2 = -1 * 9 = -9$. To avoid this potentially hazardous problem, always surround negative numbers with parentheses *before* raising them to a power.

4. **The calculator evaluates powers entered using the $\boxed{\wedge}$ key and roots entered using the $^x\sqrt{\ }$ function.**

   The $^x\sqrt{\ }$ function is found in the Math menu which is explained in Chapter 3. You can also enter various roots by using fractional exponents — for example, the cube root of 8 can be entered as $8^{\wedge}(1/3)$.

5. **The calculator evaluates all multiplication and division problems as it encounters them, proceeding from left to right.**

6. **The calculator evaluates all addition and subtraction problems as it encounters them, proceeding from left to right.**

Well, okay, what does the phrase "x plus 1 divided by x minus 2" actually *mean* when you say it aloud? Well, that depends on how you say it. Said without pausing, it means $(x + 1)/(x - 2)$. Said with a subtle pause after the "plus" and another before the "minus," it means $x + (1/x) - 2$. The calculator can't hear speech inflection, so make good use of those parentheses when you're "talking" to the calculator.

Using the Previous Answer

You can use the previous answer in the next arithmetic expression you want to evaluate. If that answer is to appear at the beginning of the arithmetic expression, first key in the operation that is to appear after the answer. The calculator displays **Ans** followed by the operation. Then key in the rest of the arithmetic expression and press $\boxed{\text{ENTER}}$ to evaluate it, as illustrated in Figure 2-3.

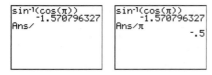

Enter first operation   Complete expression

Figure 2-3: Steps for starting an expression with the last answer.

If you want to embed the last answer in the next arithmetic expression, key in the beginning of the expression to the point where you want to insert the previous answer. Then press [2nd][(-)] to key in the last answer. Finally, key in the rest of the expression and press [ENTER] to evaluate it, as shown in Figure 2-4.

Start expression   Press [2nd] [(-)]   Complete expression

Figure 2-4: Steps for embedding the last answer in an expression.

Recycling the Last Entry

If you want to reuse the last command, function, or expression entered in the calculator — but with different instructions, arguments, or variables — you can simply recall that command, function, or expression and then edit it. To do so, follow these steps:

1. **Enter the command, function, or expression on the Home screen and then press [ENTER] to execute command or evaluate the function.**

   This is illustrated in the first picture in Figure 2-5.

2. **Press [2nd][ENTER] to paste a copy of that command or function on the home screen.**

   This procedure is shown in the second picture in Figure 2-5.

3. **Edit the command, function, or expression and then press [ENTER] to reevaluate it.**

   The third picture in Figure 2-5 shows this procedure. Editing is explained in Chapter 1.

```
sin⁻¹(cos(π))
        -1.570796327
```

```
sin⁻¹(cos(π))
        -1.570796327
sin⁻¹(cos(π))
```

```
sin⁻¹(cos(π))
        -1.570796327
sin⁻¹(cos(π/2))
                  0
```

Original expression   Press [2nd] [ENTER]   Edit expression

Figure 2-5: Steps for recycling a command, function, or expression.

Storing Variables

If you plan to use the same number many times when evaluating arithmetic expressions, consider storing that number in a variable. To do so, follow these steps:

1. **If necessary, press [2nd][MODE] to enter the Home screen.**

2. **Enter the number you want to store in a variable.**

   You can store the number as an arithmetic expression. This is illustrated in the first picture in Figure 2-6. After you complete the steps for storing the number, the calculator evaluates that expression.

3. **Press [STO▸].**

   The result of this action is shown in the first picture in Figure 2-6.

4. **Press [ALPHA] and then press the key corresponding to the letter of the variable in which you want to store the number.**

   The second picture in Figure 2-6 shows this process. The letters used for storing variables are the letters of the alphabet and the Greek letter θ.

5. **Press [ENTER] to store the value.**

   This is illustrated in the third picture in Figure 2-6.

The number you store in a variable remains stored in that variable until you *or the calculator* stores a new number in that variable. Because the calculator uses the letters X, T, and θ when graphing functions, parametric equations, and polar equations, it is possible that the calculator will change the value stored in these variables when the calculator is in graphing mode. For example, if you store a number in the variable X and then ask the calculator to find the zero of the graphed function X^2, the calculator will replace the number stored in X with 0, the zero of X^2. So avoid storing values in these three variables if you want that value to remain stored in that variable after you have graphed functions, parametric equations, or polar equations.

Press STO▸   Enter variable   Press ENTER

Figure 2-6: Steps for storing a number in a variable.

After you have stored a number in a variable, you can insert that number into an arithmetic expression. To do so, place the cursor where you want the number to appear, press ALPHA, and then press the key corresponding to the letter of the variable in which the number is stored. Figure 2-7 shows how.

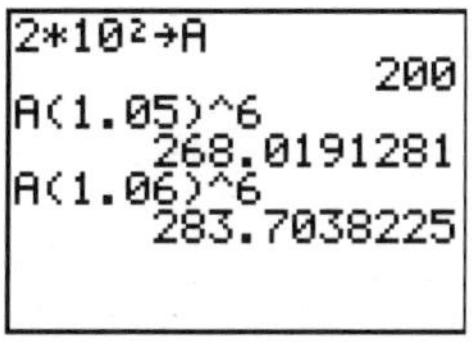

Figure 2-7: Inserting a stored variable into an expression.

Combining Expressions

You can *combine* (link) several expressions or commands into one expression by using a colon to separate the expressions or commands. The colon is entered into the calculator by pressing ALPHA . .

Combining expressions is a really handy way to write mini-programs, as detailed in the "Writing a mini-program" sidebar.

Writing a mini-program

This figure depicts a program that calculates n^2 when n is a positive integer. The first line of the program initiates the program by storing the value 1 in n and calculating its square. (Storing a number in a variable and combining expressions are explained elsewhere in this chapter.)

The next line on the left shows the real guts of the program. It increments n by 1 and then calculates the square of the new value of n. Of course, the calculator doesn't do these calculations until after you press [ENTER]. But the really neat thing about this is that the calculator will continue to execute this same command each time you [ENTER]. Because n is incremented by 1 each time you press [ENTER], you get the values of n^2 when n is a positive integer.

Part II

Advanced Functions, Complex Numbers, and Solving Equations

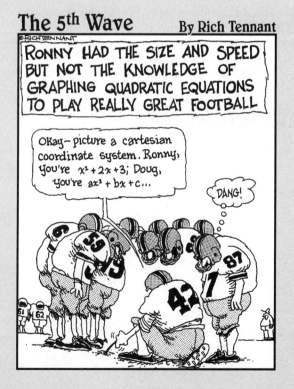

The 5th Wave   By Rich Tennant

RONNY HAD THE SIZE AND SPEED BUT NOT THE KNOWLEDGE OF GRAPHING QUADRATIC EQUATIONS TO PLAY REALLY GREAT FOOTBALL

Okay— picture a cartesian coordinate system. Ronny, you're $x^2 + 2x + 3$; Doug, you're $ax^2 + bx + c$...

DANG!

In this part...

*I*n this part, I explain how to use the numerous functions housed in the Math menu to perform tricks like converting a decimal to a fraction. Speaking of conversions, I also discuss how to convert angle measurement from degrees to radians. And if you've ever wanted a quick, powerful way to do arithmetic with complex numbers, check out the discussion of the Math CPX submenu. Finally, I show you how to solve equations using the calculator's nifty Equation Solver.

Chapter 3

The Math and Angle Menus

• •

In This Chapter

▶ Inserting a Math menu function into an expression

▶ Converting between degrees and radians

▶ Entering angles in degrees, minutes, and seconds

▶ Converting between degrees and degrees, minutes, and seconds

▶ Entering angles in degrees, minutes, and seconds

▶ Overriding the angle mode of the calculator

• •

*A*re you hunting for the absolute value function? Look no far-
ther — it's in the Math menu. Do you want to convert a decimal to a fraction? You can find the function that does this in the
Math menu. In general, any math function that cannot be directly
accessed using the keyboard is housed in the Math menu. Similarly,
any function that deals with angles is housed in the Angle menu.
This chapter tells you how to access and use those functions.

The Math Menu and Submenus

Press MATH to accesses the Math menu. This menu contains four
submenus: **MATH, NUM, CPX,** and **PRB.** Use the ▶◀ keys to get
from one submenu to the next, and back again.

The Math MATH submenu contains the general mathematical functions such as the cubed root function. It also contains the calculator's Equation Solver that, as you would expect, is used to solve
equations. The Equation Solver is explained in Chapter 5. The Math
NUM submenu contains the functions usually associated with numbers, such as the least common multiple function. A detailed explanation of the functions in these two menus is given later in this
chapter.

The Math CPX submenu contains functions normally used with complex numbers. This submenu is explained in detail in the next chapter. The Math PRB submenu contains the probability and random-number functions. (Probability is explained in Chapter 17.)

Using Math menu functions

Most functions housed in the Math menu require that the argument be entered after entering the function. Such functions are easily recognizable by the parenthesis following the name of the function; the cube-root function, ³√(, is an example.

With one exception, functions in the Math menu that have no parenthesis at the end of their names require that the argument be entered before the function is entered. The cube function is an example of such a function. (The exception is the xth root function, ˣ√. This function requires that the root be entered first, then the ˣ√ function, and then the argument.)

Inserting a Math menu function

Math menu functions are usually inserted into arithmetic expressions entered on the Home screen, or into definitions of functions in the Y= editor. Pressing [2nd][MODE] puts you in the Home screen, and pressing [Y=] puts you in the Y= editor. The Y= editor is used to define functions you want to graph. This editor is explained in detail in Chapters 9, 13, 15, and 16. To insert a Math menu function into an expression, follow these steps:

1. **If necessary, use [▶][◀][▲][▼] to place the cursor where you want to insert the function.**

   Inserting a Math menu item in an arithmetic expression that is entered on the Home screen is illustrated in Figure 3-1. This figure shows how to insert the cube function in order to evaluate √(2³ + 17).

2. **Press [MATH].**

3. **Use [▶] to select the appropriate submenu of the Math menu.**

4. **Enter the number of the function or use [▼] to highlight the number of the function and then press [ENTER].**

5. **If necessary, complete the expression.**

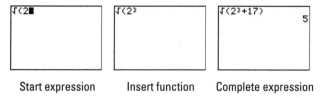

Start expression   Insert function   Complete expression

Figure 3-1: Steps for inserting a Math menu function in an expression.

The Math MATH submenu

Press [MATH] to access the Math MATH submenu. This submenu contains general mathematical functions you can insert into an expression. The following sections explain the items housed in this submenu, except for the **Solver** function at the bottom of the Math MATH submenu. This latter function, used to solve equations, is discussed in Chapter 5.

Converting between decimals and fractions

The **Frac** function always converts a finite decimal to a fraction. Sometimes this function can convert an infinite repeating decimal to a fraction. When it can't, it lets you know by redisplaying the decimal. Be sure to enter the decimal before inserting the **Frac** function, as shown in the first two lines of the first picture in Figure 3-2.

The **Dec** function converts a fraction to a decimal. Enter the fraction before you insert the **Dec** function. An example is shown in the first picture in Figure 3-2.

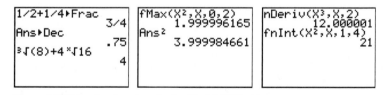

Figure 3-2: Using the functions in the Math MATH submenu.

Cubing and taking cube roots

The cube function, ^{3}, cubes the value that precedes the function (as illustrated in Figure 3-1). The cube-root function, $^3\sqrt{}$, finds the cube root of a value that follows the function. This value must be enclosed in parentheses; the calculator supplies the opening parenthesis; you must supply the closing parenthesis. The first picture in Figure 3-2 shows an example of using the cube-root function.

Taking the xth root

The xth root function, $\sqrt[x]{}$, finds the xth root of the value that follows the function. To use this function, first enter the root x, then insert the $\sqrt[x]{}$ function, and then enter the argument. This is illustrated in the first picture in Figure 3-2. In this example the calculator is evaluating the fourth root of 16.

Finding the location of maximum and minimum values

The **fMin** and **fMax** functions approximate *where* the minimum or maximum value of a function occurs in a specified interval. *They do not compute the minimum or maximum value of the function;* they just give you the x-coordinate of the minimum or maximum point. Chapter 11 tells you how to get the calculator to compute minimum and maximum values of a function.

The **fMin** and **fMax** functions are standalone functions in the sense that they cannot be used in an expression. To use these functions, insert the appropriate function, **fMin** or **fMax,** at the beginning of a new line on the Home screen. Then enter the definition of the function whose minimum or maximum you want to locate. Press ⎡,⎤ and enter the variable used in the definition of the function you just entered. Press ⎡,⎤ and enter the lower limit of the specified interval. Then press ⎡,⎤, enter the upper limit, and press ⎡)⎤. Finally, press ⎡ENTER⎤ to *approximate* the location of the minimum or maximum in the specified interval. This is illustrated in the second picture in Figure 3-2. In this picture the calculator is *approximating* the location of the maximum value of the function x^2 in the interval $0 \le x \le 2$.

Doing numerical differentiation and integration

The calculator cannot perform symbolic differentiation and integration. For example, it cannot tell you that the derivative of x^2 is $2x$, nor can it evaluate an indefinite integral. But the **nDeriv** function will approximate the derivative (slope) of a function at a specified value of the variable, and the **fnInt** function will approximate a definite integral.

Insert the **nDeriv** function, enter the function whose derivative you want to find, and then press ⎡,⎤. Enter the variable used in the definition of the function you just entered and press ⎡,⎤. Then enter the value at which the derivative is to be taken, and press ⎡)⎤. Finally, press ⎡ENTER⎤ to *approximate* the derivative. This is illustrated in the third picture in Figure 3-2.

To use the **fnInt** function, insert **fnInt**, and then enter the function you are integrating. Press ⎡,⎤ and enter the variable used in the definition of the function you just entered. Press ⊡ and enter the lower limit of the definite integral. Press ⊡, enter the upper limit, and press ⊡. Finally, press [ENTER] to *approximate* the definite integral. This is illustrated in the third picture in Figure 3-2.

The calculator may give you an error message or a false answer if **nDeriv** is used to find the derivative at a nondifferentiable point or if **fnInt** is used to evaluate an improper integral.

The Math NUM submenu

Press [MATH]⊡ to access the Math NUM submenu. (Inserting a Math menu function into an expression is explained earlier in this chapter.) The following sections explain the items housed in this submenu:

Finding the absolute value

The **abs** function evaluates the absolute value of the number or arithmetic expression that follows the function. This number or expression must be enclosed in parentheses. The calculator supplies the first parenthesis; you must supply the last parenthesis. An example of using the **abs** function is illustrated in the first picture in Figure 3-3.

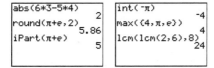

Figure 3-3: Using the functions in the Math NUM submenu.

Rounding numbers

The **round** function rounds a number or arithmetic expression to a specified number of decimal places. The number or expression to be rounded and the specified number of decimal places are placed after the function separated by a comma and surrounded by parentheses. The calculator supplies the opening parenthesis; you must supply the closing parenthesis. An example of using the **round** function is the first picture in Figure 3-3.

Finding the integer and fractional parts of a value

The **iPart** and **fPart** functions (respectively) find the integer and fractional parts of the number, or the arithmetic expression that follows the function. This number or expression must be enclosed in parentheses. The calculator supplies the opening parenthesis; you must supply the closing parenthesis. An example of using the **iPart** function is the first picture in Figure 3-3.

Using the greatest-integer function

The **int** function finds the largest integer that is less than or equal to the number or arithmetic expression that follows the function. This number or expression must be enclosed in parentheses. The calculator supplies the opening parenthesis; you must supply the closing parenthesis. An example of using the **int** function is the second picture in Figure 3-3.

Finding minimum and maximum values in a list of numbers

The **min** and **max** functions find (respectively) the minimum and maximum values in the list of numbers that follows the function. Braces must surround the list, and commas must separate the elements in the list. You can access the braces on the calculator by pressing 2nd (and 2nd). The list must be enclosed in parentheses. The calculator supplies the opening parenthesis; you must supply the closing parenthesis. An example of using the **max** function is the second picture in Figure 3-3.

When using the **min** or **max** function to find the minimum or maximum of a two-element list, you can omit the braces that surround the list. For example, **min**(2, 4) returns the value 2.

Finding least common multiple and greatest common divisor

The **lcm** and **gcd** functions find (respectively) the least common multiple and the greatest common divisor of the two numbers that follow the function. These two numbers must be separated by a comma and surrounded by parentheses. The calculator supplies the opening parenthesis; you must supply the closing parenthesis. An example of using the **lcm** function is the second picture in Figure 3-3.

The Angle Menu

The functions housed in the Angle menu allow you to convert between degrees and radians or convert between rectangular and polar coordinates. They also allow you to convert between degrees and DMS (degrees, minutes, and seconds). You can also override the angle setting in the Mode menu of the calculator when you use these functions. For example, if the calculator is in Radian mode and you want to enter an angle measure in degrees, there is a function in the Angle menu that allows you to do so. (Setting the mode is explained in Chapter 1.)

Converting between rectangular and polar coordinates is explained in Chapter 16. The other topics are explained in the next sections.

Converting degrees to radians

To convert degrees to radians, follow these steps:

1. **Put the calculator in Radian mode.**

   Setting the mode is explained in Chapter 1.

2. **If necessary, press [2nd][MODE] to access the Home screen.**

3. **Enter the number of degrees.**

4. **Press [2nd][APPS][1] to paste in the ° function. (On the TI-83, press [2nd][MATRX][1].)**

5. **Press [ENTER] to convert the degree measure to radians.**

   This is illustrated at the top of the first picture in Figure 3-4.

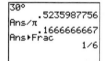
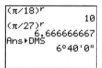

Figure 3-4: Converting between degrees and radians.

If you're a purist (like me) who likes to see radian measures expressed as a fractional multiple of π whenever possible, continuing with the following steps accomplishes this goal if it's mathematically possible.

6. **Press [÷][2nd][^][ENTER] to divide the radian measure by π.**

   This is illustrated in the third line in the first picture in Figure 3-4.

7. **Press [MATH][1][ENTER] to convert the result to a fraction, if possible.**

   This is illustrated at the bottom of the first picture in Figure 3-4, where 30° is equal to π/6 radians. If the calculator can't convert the decimal obtained in Step 6 to a fraction, it says so by returning the decimal in Step 7.

Converting radians to degrees

To convert radians to degrees:

1. **Put the calculator in Degree mode.**

   Setting the mode is explained in Chapter 1.

2. **If necessary, press 2nd MODE to access the Home screen.**

3. **Enter the radian measure.**

   If the radian measure is entered as an arithmetic expression, surround that expression with parentheses. This is illustrated at the top of the second picture in Figure 3-4.

4. **Press 2nd APPS 3 to paste in the *r* function. (On the TI-83, press 2nd APPS 3 .)**

5. **Press ENTER to convert the radian measure to degrees.**

   This is illustrated in the second picture in Figure 3-4.

Converting between degrees and DMS

To convert degrees to DMS (degrees, minutes, seconds), follow these steps:

1. **Put the calculator in Degree mode.**

   Setting the mode is explained in Chapter 1.

2. **If necessary, press 2nd MODE to access the Home screen.**

3. **Enter the degree measure.**

4. **Press 2nd APPS 4 ENTER to convert the degrees to DMS. (On the TI-83, press 2nd MATRX 4 ENTER .)**

   This is illustrated at the bottom of the second picture in Figure 3-4, and also at the top of Figure 3-5.

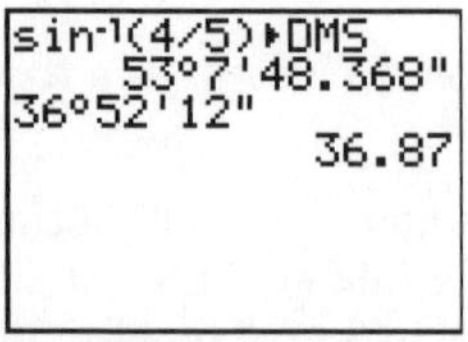

```
sin⁻¹(4/5)▶DMS
        53°7'48.368"
36°52'12"
                36.87
```

Figure 3-5: Converting between degrees and DMS.

To convert DMS to degrees, follow these steps:

1. **Follow the above Steps 1 and 2.**
2. **Enter the DMS measure.**

   The next section tells you how to insert the symbols for degrees, minutes, and seconds.
3. **Press [ENTER] to convert the DMS entry to degrees.**

   This is illustrated at the bottom of Figure 3-5.

Entering angles in DMS measure

To enter an angle in DMS measure:

1. **Enter the number of degrees and press [2nd][APPS][1] to insert the degree symbol. (On the TI-83, press [2nd][MATRX][1].)**
2. **Enter the number of minutes and press [2nd][APPS][2] to insert the symbol for minutes. (On the TI-83, press [2nd][MATRX][2].)**
3. **Enter the number of seconds and press [ALPHA][+] to insert the symbol for seconds.**

   This is illustrated at the bottom of Figure 3-5.

Overriding the mode of the angle

If the calculator is in Radian mode but you want to enter an angle in degrees, enter the number of degrees and then press [2nd][APPS][1] to insert in the ° degree symbol. (On the TI-83, press [2nd][MATRX][1].) If you want to enter an angle in DMS measure while the calculator is in Radian mode, the previous section tells you how to do so. (Setting the mode is explained in Chapter 1.)

If the calculator is in Degree mode and you want to enter an angle in radian measure, enter the number of radians and then press [2nd][APPS][3] to insert in the radian-measure symbol. (On the TI-83, press [2nd][MATRX][3].) If the radian measure is entered as an arithmetical expression, such as $\pi/4$, be sure to surround it with parentheses before you insert the radian-measure symbol.

Chapter 4

Dealing with Complex Numbers

· ·

In This Chapter

▶ Setting the Mode menu for complex numbers

▶ Entering (and doing basic arithmetic with) complex numbers

▶ Examining the functions housed in the Math CPX submenu

· ·

*A*ll the arithmetic operations described in Chapter 2 apply to complex numbers, provided the calculator is in Complex mode. This mode is explained in this chapter, as well as how to enter and do basic arithmetic with those complex numbers. The Math menu, described in Chapter 3, has a submenu devoted to complex numbers. How to use this submenu is also described in this chapter.

Setting the Mode

The Mode menu is pictured in Figure 4-1. The second from the last line of the Mode menu gives you three options: **Real, a + b***i,* and **re^θ***i.* The last two options are used when working with complex numbers. The **a + b***i* options tells the calculator to display complex numbers in rectangular form, and the **re^θ***i* options tells it to display complex numbers in polar form.

Figure 4-1: Putting the calculator in Complex mode.

Select either **a + b***i* or **re^θ***i* so the calculator knows you're working with complex numbers. The remaining menu items can be set as they appear in Figure 4-1. (Setting the Mode, as well as an explanation of the other options in this menu, can be found in Chapter 1.)

Using Complex Numbers

Independent of the mode setting, **a + b***i* or **re^θ***i*, complex numbers can be entered in the calculator in either *a + bi* rectangular form or in *re^θi* polar form. You enter the imaginary number *i* in an expression by pressing 2nd⊡; *e* is entered by pressing 2nd÷. Figures 4-2 and 4-5 show complex numbers entered (respectively) in rectangular form and in polar form.

If the calculator is in Radian mode, θ in the polar form of a complex number is entered in radian measure; if the calculator is in Degree mode, θ is entered in degrees. (The last section in Chapter 3 tells you how to override the restriction on entering θ in degrees so you can do so even when the calculator is in Radian mode.)

Chapter 2 tells you how to evaluate arithmetic expressions. Although the figures in this chapter involve real-number arithmetic, everything in that chapter can be used when doing complex-number arithmetic — provided the calculator is in Complex mode, as described in the previous section. An example of this appears in Figure 4-2. (Remember, you can enter the imaginary number *i* into an expression by pressing 2nd⊡.)

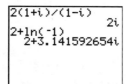

Figure 4-2: Doing arithmetic with complex numbers.

The Math CPX Submenu

Press MATH▶▶ to access the Math CPX submenu. This submenu contains the standard functions used with complex numbers. Inserting a Math menu function into an expression is explained in Chapter 3. The following sections explain the items housed in this submenu.

Finding the conjugate of a complex number

The **conj** function evaluates the conjugate of the complex number or arithmetic expression that follows the function. This number or expression must be enclosed in parentheses. The calculator supplies the opening parenthesis; you must supply the closing parenthesis. An example of using the **conj** function is illustrated in Figure 4-3.

Finding the real or imaginary part of a complex number

The **real** and **imag** functions find (respectively) the real and imaginary parts of any complex number or arithmetic expression that follows the function. This number or expression must be enclosed in parentheses. The calculator supplies the opening parenthesis; you must supply the closing parenthesis. An example of using the **imag** function is illustrated in Figure 4-3.

```
(1+i)-conj(2+3i)
                -1+4i
imag(2+ln(-1))
        3.141592654
```

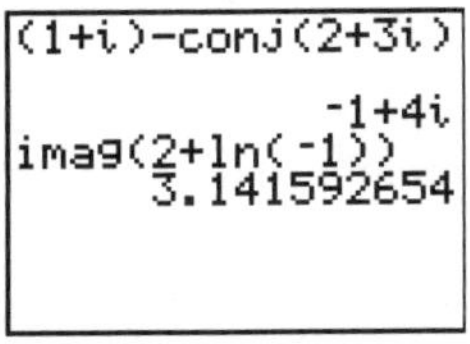

Figure 4-3: Finding the conjugate and the imaginary part.

Finding the polar angle and modulus (magnitude) of a complex number

The **angle** and **abs** functions respectively find the polar angle and the magnitude (modulus) of the complex number or arithmetic expression that follows the function. This number or expression must be enclosed in parentheses. The calculator supplies the opening parenthesis; you must supply the closing parenthesis. An example of using these functions is illustrated in Figure 4-4.

The **angle** function gives the polar angle in radians if the calculator is in Radian mode, or in degrees if the calculator is in Degree mode. (Chapter 3 tells you how to convert between radians and degrees, express a radian measure as a fractional multiple of π, and convert degrees to degrees, minutes, and seconds.)

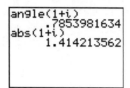

Figure 4-4: Finding the polar angle and modulus.

Converting between rectangular and polar form

The **Rect** function converts a complex number from polar form to rectangular from. The polar form of the complex number must come before the **Rect** function, as shown in the first picture in Figure 4-5.

The **Polar** function converts a complex number from rectangular form to polar from. The rectangular form of the complex number must come before the **Polar** function. This is illustrated in the first picture in Figure 4-5. This function won't convert a real number to polar form; instead, it returns the ERR: DATA TYPE error message.

If the answer produced by employing either of these functions doesn't fit on the screen (as indicated by the ellipsis in the last line of the first picture in Figure 4-5), repeatedly press ▶ to display the rest of the answer, as shown in Figure 4-5.

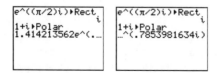

Figure 4-5: Converting between rectangular and polar form.

Chapter 5

Solving Equations

● ●

In This Chapter

▶ Entering, editing, and solving equations in the Equation Solver

▶ Assigning values to the variables in the equation

▶ Defining the interval containing the solution

▶ Guessing the value of the solution

▶ Finding multiple solutions

● ●

The Equation Solver solves an equation for one variable given the values of the other variables in the equation. The Solver is capable of dealing only with real numbers and is capable of finding only real-number solutions.

Using the Equation Solver

The following lists the basic steps for using the Equation Solver. Each of these steps is explained in detail below this list. If you have never used the Equation Solver before, I suggest that you read the detailed explanations for each step because the Equation Solver is a bit tricky. After you have had experience using the Solver, you can refer back to this list, if necessary, to refresh your memory on its use.

1. **If necessary, set the Mode for real-number arithmetic.**

2. **Enter a new equation (or edit the existing equation) in the Equation Solver.**

3. **Assign values to all variables except the variable you're solving for.**

4. **Enter the bounds for the interval that contains the solution.**

5. **Enter a guess for the solution.**

6. **Press** ALPHA ENTER **to solve the equation.**

Step 1. Set the mode

Because the Equation Solver is equipped to deal only with real numbers, press MODE and highlight all entries on the left. (To see what I mean, refer to Figure 1-3 in Chapter 1).

Step 2. Enter or edit the equation to be solved

To enter a new equation in the Equation Solver, follow these steps:

1. **Press MATH 0 to access the Equation Solver from the Math menu.**

   When the Equation Solver appears, it looks like one of the pictures in Figure 5-1. The first picture shows the Equation Solver when no equation is stored in the Solver; the second picture depicts the Solver and the equation currently stored in it.

Without an equation   With an equation

Figure 5-1: The Equation Solver.

2. **If your Equation Solver already contains an equation, repeatedly press ▲ until the screen titled EQUATION SOLVER appears.**

   This is illustrated in the first picture in Figure 5-2.

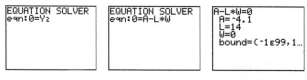

Press ▲   Enter equation   Press ENTER

Figure 5-2: Steps for entering a new equation in the Equation Solver.

3. **If you need to, press CLEAR to erase any equation in the Solver.**

4. **Enter the equation you want to solve.**

The equation you enter must be set equal to zero and can contain only real numbers. For example, if you want to solve the area equation A = L*W, enter it as 0 = A – L*W or as 0 = L*W – A (as illustrated in the second picture in Figure 5-2).

You can also use a function that you've entered in the Y= editor in the definition of your equation. For example, if $Y_1 = X - 1$ in the Y= editor, you can enter $1/Y_1 - 1$ in the Equation Solver to solve the equation $1/(x - 1) - 1 = 0$. To insert such a function into the equation, press [VARS][▶][1] to access the Function menu, and then press the number of the function you want to enter (as in the first picture in Figure 5-2). The Y= editor is explained in Chapter 9.

5. **Press [ENTER] to enter the equation in the solver.**

To edit an equation that is already entered in the Equation Solver, follow these steps:

1. **Follow the above Steps 1 and 2.**

2. **Edit the equation and press [ENTER] when you're finished.**

Step 3. Assign values to variables

After you have entered an equation in the Equation Solver, the values assigned to the variables in your equation are the values that are currently stored in those variables in your calculator. This is illustrated in the last picture in Figure 5-2. You must assign an accurate value to all variables except the variable you're solving for. These values must be real numbers or arithmetic expressions that simplify to real numbers.

To assign a value to a variable, use [▶][◀][▲][▼] to place the cursor on the number currently assigned to that variable and then key in the new value. As you start to key in the new value, the old value is erased. Press [ENTER] when you're finished entering the new value (as illustrated in the first picture of Figure 5-3, where values are assigned to variables **A** and **W**).

Step 4. Define the solution bounds

The **bound** variable at the bottom of the screen, as illustrated in the first picture in Figure 5-3, is where you enter the bounds of the interval containing the solution you're seeking. The default setting for this interval is $[-10^{99}, 10^{99}]$, as is indicated by **bound** = {-1E99, 1E99}. 1E99 is 10^{99} in scientific notation. The ellipsis at the end of the line containing this variable indicate that you have to repeatedly press ▶ to see the rest of the line.

This default setting is more than sufficient for equations that have a unique solution. So if your equation has a unique solution, you don't have to do anything with the value in the **bound** variable.

When the equation you're solving has multiple solutions, it's sometimes necessary to redefine the **bound** variable. Finding multiple solutions is discussed in the last section of this chapter.

To redefine the **bound** variable:

1. **Use ▶ ◀ ▲ ▼ to place the cursor anywhere in the line containing the bound variable.**

2. **Press CLEAR to erase the current entry.**

3. **Press 2nd (to insert the left brace.**

4. **Enter the lower bound, press ․, enter the upper bound, and then press 2nd) to insert the right brace.**

5. **Press ENTER to store the new setting in the bound variable.**

Step 5. Guess a solution

Guess at a solution by assigning a value to the variable you're solving for. Any value in the interval defined by the **bound** variable will do. If your guess is close to the solution, the calculator quickly solves the equation; if it's not, it may take the calculator a while to solve the equation. (Assigning a value to a variable in the Equation Solver is explained earlier in this chapter.)

If your equation has more than one solution, the calculator will find the one closest to your guess. The section at the end of this chapter tells you how to find the other solutions.

If the variable you're solving for is assigned a value (guess) that isn't in the interval defined by the **bound** variable, then you get the ERR: BAD GUESS error message.

Step 6. Solve the equation

To solve an equation, follow these steps:

1. **Use ▶◀▲▼ to place the cursor anywhere in the line that contains the variable you're solving for.**

    This procedure is shown in the second picture in Figure 5-3.

2. **Press ALPHA ENTER to solve the equation.**

    The third picture of Figure 5-3 shows this procedure; the square indicator shown next to the **L** indicates that **L** is the variable just solved for. The **left – rt** value that appears at the bottom of this picture evaluates the two sides of the equation (using the values assigned to the variables) and displays the difference — that is, the accuracy of this solution. A **left – rt** value of zero indicates an exact solution. Figure 5-4 shows a solution that is off by the very small number $-1 * 10^{-11}$.

If you get the ERR: NO SIGN CHNG error message when you attempt to solve an equation using the Equation Solver, then the equation has no real solutions in the interval defined by the **bound** variable.

Define variables	Guess solution	Press ALPHA ENTER

Figure 5-3: Steps for solving an equation in the Equation Solver.

Finding Multiple Solutions

To find other solutions to an equation, first find one solution to the equation by following Steps 1 through 6 in the first section of this chapter. This is illustrated in the first picture in Figure 5-4.

Then enter a new guess for the solution you're seeking, or, in the **bound** variable, enter the bounds of an interval that possibly contains a different solution. In the second picture in Figure 5-4 a new guess for the solution is entered.

After making a new guess or after redefining the **bound** variable, follow the steps in the previous section to find another solution to the equation. The third picture in Figure 5-4 shows this procedure.

Find 1st solution   Enter new guess   Press ALPHA ENTER

Figure 5-4: Steps for finding multiple solutions to an equation.

Part III
Dealing with Finances

In this part...

This part explains how to use the financial features on the calculator to answer many important questions — which run the gamut from "Should I lease or borrow?" to "How much should I invest if I want to retire as a millionaire?" I also discuss how to calculate the best interest rate, find internal rates of return, use the (Time-Value-of-Money) TVM Solver, and cope with round-off errors.

Chapter 6

Finding the Best Deal

. . . .

In This Chapter

▶ Finding the best interest rate

▶ Converting between nominal and effective rates

▶ Deciding whether to lease or to take out a loan

▶ Finding the internal rate of return

. . . .

Finding the Best Interest Rate

Which of the following is the best interest rate for a savings account?

✔ 5.120% annual rate, compounded monthly

✔ 5.116% annual rate, compounded daily

✔ 5.115% annual rate, compounded continuously

When *nominal rates* (also called *annual percentage rates*) are compounded at different frequencies (as are those just given), you can compare them to each other only by converting them to *effective rates* (the simple-interest equivalent of nominal rates).

Finding the effective rate

To find the effective rate given the nominal rate:

1. **Set the second line of the Mode menu to Float.**

   When dealing with money, you usually set the second line of the Mode menu to **2** so all numbers are rounded to two decimal places. When you're comparing interest rates, however, you want to see as many decimal places as possible. You can do so by setting the second line to **Float**. (For more about setting the Mode menu, see Chapter 1.)

2. **Press** [APPS][1] **to start the Finance application. (On the TI-83, press** [2nd][x⁻¹]**.)**

3. **Repeatedly press** [▾] **to move the indicator to the Eff command and press** [ENTER]**.**

4. **Enter the nominal rate, press** [.]**, enter the number of compounding periods per year, and press** [ )][ENTER]**.**

When interest is compounded continuously, it is compounded an infinite number of times a year. But there is no way of entering infinity into the calculator. You can get around this problem by entering a very large number, such as 10^{12}, for the number of compounding periods. The fastest way to enter this number is to press [2nd][.] and then enter the number **12**.

Figure 6-1 illustrates this procedure; it also shows that the answer to the question posed at the beginning of this section is 5.116%, compounded daily (the choice that gives you the largest effective rate).

You can easily use the same command over and over (as in Figure 6-1), if (after using the command the first time) you press [2nd][ENTER] to recopy the command to the next line on the screen. Then edit the entries in the command and press [ENTER] to execute the command. (Editing is explained in Chapter 1.)

```
►Eff(5.120,12)
         5.241874641
►Eff(5.116,365)
         5.248750517
►Eff(5.115,E12)
         5.248075356
```

Figure 6-1: Finding the effective rate.

Finding the nominal rate

The steps that convert an effective rate to a nominal rate are similar to those listed in the previous section. To begin, you follow the first two of those steps — and then (in Step 3), select the **Nom** command instead of the **Eff** command. In Step 4, you enter the effective rate (after the comma, enter the number of compounding periods for the nominal rate).

Leasing versus Borrowing

Suppose you're planning to purchase a $2,000 laptop. Which of the following approaches can give you the better deal?

- ✔ Lease the laptop for $600 a year for four years with the option to buy the laptop after four years for an additional $300.

- ✔ Take out a four-year loan at 12% simple interest.

The *internal rate of return* is the yearly simple interest rate that you earn on an investment plan. In the context of a lease, the internal rate of return is the yearly simple interest rate you would pay if the lease were converted to a loan. So to find the better deal, you must compare the lease's internal rate of return to 12% (the loan's internal rate of return).

To find the lease's internal rate of return:

1. **Select the** irr **command from the Finance application menu.**

   To do so, follow the first two steps in the previous section ("Finding the Effective Rate") — but in Step 3, select the **irr** command instead of the **Eff** command.

2. **Enter the initial cash flow and press ⊡.**

   For the leasing program described at the beginning of this section (for example), the initial cash flow is the $2,000 price of the laptop. The sidebar in at the end of this chapter explains why this value is positive even though the cash is flowing away from you.

3. **Enter the cash-flow list and press ⊡.**

   In the leasing-program scenario, the cash flow indicates what you paid per year to lease the laptop: $600 a year for the first three years, and $900 in the fourth year (that is, $600 to lease it and then $300 to purchase it). So the cash-flow list is {-600, -900}. (Don't worry: The sidebar at the end of this chapter explains why these values must be negative.) You enter the frequencies for this list in the next step.

   Enter the cash-flow list as a list contained within braces, using commas to separate the elements in the list (as in Figure 6-2). You enter the braces into the calculator by pressing ⟨2nd⟩⟨(⟩ and ⟨2nd⟩⟨)⟩. (Remember to use the ⟨(-)⟩ key to indicate that a number is negative.)

4. Enter the cash-flow frequency list.

The *cash-flow frequency list* indicates how frequently each element (in this case, each specific amount) occurs in the cash-flow list. In the leasing-program scenario, the cash-flow frequency list is {3, 1} because $600 was paid for the first three years and $900 was paid in the fourth year. Figure 6-2 shows this case.

5. Press ⬚ [ENTER] to calculate the internal rate of return.

This procedure is illustrated in Figure 6-2. This figure also shows the answer to the question posed at the beginning of this section: You're better off taking out the loan at 12% simple interest because leasing the laptop is equivalent to a 12.29% loan.

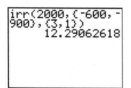

```
irr(2000,{-600,-
900},{3,1})
         12.29062618
```

Figure 6-2: Finding the internal rate of return.

When is cash flow negative?

Cash flow is the money that changes hands. The calculator does not know who is involved in this exchange, so it requires that you indicate which way the money is going by using a positive or negative value for cash flow. Figuring out which sign you should use isn't always easy.

The best way to figure out which sign to use is to ask yourself whether the money is going into your pocket or coming out of your pocket. If it is going *into* your pocket, then you have more, so the cash flow is positive. If it is coming *out of* your pocket, then you have less, so it's negative.

For example, if you put $2,000 in a savings account, that's a negative cash flow because it came out of your pocket and went to the bank. It's still your money, but it's no longer in your pocket.

On the other hand, if you take out a $2,000 loan, then that's a positive cash flow because the bank has given you the money. In your mind, it's negative because you have to pay it off. But to the calculator, it's positive because now you have the money.

Chapter 7

Loans and Mortgages

. . . .

In This Chapter

▶ Using the Time-Value-of-Money (TVM) Solver

▶ Using a TVM value in a calculation

▶ Finding the principal paid on a loan during a specified time period

▶ Finding the interest paid on a loan during a specified time period

▶ Finding the balance of a loan after a specified time period

. . . .

*Y*ou have a 30-year, $200,000 mortgage on your house. The mortgage rate is 7%, compounded monthly.

✔ What are your monthly mortgage payments?

✔ What is the total cost of the loan?

✔ How much of your first payment is devoted to paying off the balance of the loan?

✔ How much of the loan was paid off during the second year?

✔ How much interest do you pay during the life of the loan?

✔ How much do you still owe on the house after 20 years?

This chapter shows you how to get the calculator to answer these and other, similar questions.

Using the TVM Solver

The **TVM** (time-value-of-money) **Solver** can be used to answer questions like those posed at the beginning of this chapter. In fact, if you tell the **TVM Solver** any four of the following five variables, it will figure out the fifth variable for you:

▐ ✔ **N:** Total number of payments

▐ ✔ **%:** Annual interest rate

✔ **PV:** Present value

✔ **PMT:** Amount of each payment

✔ **FV:** Future value

To get the calculator to find these answers, follow these steps:

1. **Set the second line in the Mode menu to 2.**

   This setting makes the calculator round all numbers to two decimal places, the standard format for money. (Setting the Mode menu is described in Chapter 1.)

2. **Press** [APPS][1][1] **to select the TVM Solver from the Finance application menu. (On the TI-83, press** [2nd][x⁻¹][1]**.)**

3. **Enter values for four of the first five variables listed in the TVM Solver. Press** [ENTER] **after making each entry.**

   Some values that are entered in the **TVM Solver** must be entered as negative numbers. For an explanation of when you have to do this, see Chapter 6.

   This step is illustrated in the first picture in Figure 7-1. In this figure, the **TVM Solver** is set up to solve the first question asked at the beginning of this chapter.

   Don't worry about any value currently assigned by the calculator to the variable that the **TVM Solver** is going to find for you. In this example, that variable is **PMT**, the monthly payment.

   For a loan, the present value is always the amount of the loan; the future value, after the loan is paid off, is (naturally) 0.

   You can enter arithmetic problems as values in the **TVM Solver**. The calculator will do the arithmetic after you press [ENTER]. For example, in the first picture in Figure 7-1, **N** was entered as 30*12.

4. **Enter values for P/Y and C/Y. Press** [ENTER] **after making each entry.**

   • **P/Y** is the number of payments made each year.

   • **C/Y** is the number of times interest is compounded each year.

The calculator assumes these two values are the same. So after you enter the value for **P/Y** and press ENTER, the same value is assigned to the variable **C/Y**. (If the values are *not* the same in the scenario for which you're using the **TVM Solver**, enter the correct value for **C/Y**. As an example, if you're making monthly payments but the interest rate is a simple interest rate, then **P/Y** = 12 and **C/Y** = 1.)

If interest is compounded continuously, it is compounded an infinite number of times a year. But there is no way of entering infinity into the calculator. You can get around this problem by setting **C/Y** equal to a large number, such as 10^{12}.

5. In the last line, indicate whether payments are made at the beginning or the end of the payment period.

To do so, use ▶◀ to place the cursor on the appropriate entry, either **END** or **BEGIN**, and press ENTER.

6. Place the cursor on the variable for which you want to solve.

This operation is shown in the second picture in Figure 7-1. In this figure the calculator is told to solve for **PMT**, the monthly payment.

7. Press ALPHA ENTER **to solve for the variable.**

The calculator places a square next to the variable it just found. This is illustrated in the third picture in Figure 7-1.

See the sidebar in Chapter 6 for an explanation of why **PMT** is negative in the last picture in Figure 7-1.

8. Press 2nd MODE **to exit (quit) the TVM Solver and return to the Home screen.**

Define variables   Select unknown   Press ALPHA ENTER
                      variable

Figure 7-1: Steps for using the TVM Solver.

Using a TVM Value

After you've assigned values to all variables in the **TVM Solver** (as explained in the previous section), you can paste these values into other expressions by pressing APPS 1 ▶ and entering the number of the variable you want to use. (On the TI-83, press 2nd x⁻¹ ▶.)

An example of this use appears in Figure 7-2. Also, this figure answers the question (*What is the total cost of the loan?*) posed at the beginning of this chapter. Notice that this calculation was done on the Home screen. To get from the **TVM Solver** to the Home screen, press 2nd MODE.

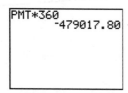

```
PMT*360
         -479017.80
```

Figure 7-2: Using a TVM variable in a calculation.

Finding Principal and Interest

Use the ΣPrn and ΣInt commands to compute the principal and interest paid on a loan during a specified time period. To use these commands, follow these steps:

1. **Assign values to all variables in the TVM Solver.**

   This chapter's "Using the TVM Solver" section tells you how.

2. **Press 2nd MODE to enter the Home screen.**

3. **Press APPS 1 to access the Finance application menu. (On the TI-83, press 2nd x⁻¹.)**

4. **Repeatedly press ⏷ to select ΣPrn or ΣInt and press ENTER.**

5. **Enter the number of the first payment in the specified time period, press ., enter the number of the last payment in the time period, press), and then press ENTER to display the result.**

This process is illustrated in Figure 7-3 (where you can also see answers to the third, fourth, and fifth questions posed at the beginning of this chapter).

```
ΣPrn(1,1)
            -163.93
ΣPrn(13,24)
           -2178.42
ΣInt(1,360)
         -279021.94
```

Figure 7-3: Evaluating cumulative principal and interest payments.

Finding the Balance

Use the **bal** command to compute the balance of a loan after a specified time period. To use this command, follow these steps:

1. **Follow Steps 1 through 3 in the previous section.**

2. **Repeatedly press ⊡ to select the bal command and press** ENTER.

3. **Enter the number of payments in the specified time period.**

   You may find the "How Many Days Till Christmas?" sidebar in the next chapter useful for calculating the number of payments in a specified time period.

   You can enter the number of payments in the **bal** command as an arithmetic expression, as in Figure 7-4.

4. **Press ⬚ENTER to display the result.**

   This procedure is shown in Figure 7-4. (This figure answers the last question posed at the beginning of this chapter.)

```
bal(20*12)
          114602.70
```

Figure 7-4: Finding the balance of a loan.

Understanding round-off errors

Do you see a problem with the results the calculator gave in Figures 7-2 and 7-3? According to Figure 7-2, the total cost of the house in the scenario posed at the beginning of this chapter is $479,017.80. According to Figure 7-3, however, the total cost is $479,021.94 (total interest plus the $200,000 mortgage). That's a difference of $4.14! The differences in these calculations are caused by 30 years worth of *round-off errors*.

In Figure 7-1, the calculator rounded the monthly payments to $1,330.60. But if you had put the calculator in Float mode, the calculator would have told you that the monthly payments are $1,330.60499. When the calculator uses this value to evaluate other entities, it uses an even better approximation than this one. But it's still using an approximation, so expect round-off errors.

The mortgage lender would avoid this problem by charging $1,330.61 a month for all but the last month.

Chapter 8

Savings and Investments

. . . .

In This Chapter

▶ Finding how long it takes to reach a financial goal

▶ Finding the future value of money

▶ Finding the present value of money

▶ Calculating the number of days between two dates

. . . .

Say you're investing money in an account that earns 6% interest, compounded continuously.

✔ If you invest $50,000 now and then add $1,000 to the account each month, how many years will it take for the account to reach $1,000,000?

✔ If you invest $50,000 now and then add $1,000 to the account each month, how much money will you have after 30 years?

✔ How much should you invest now so that the account will be worth at least $1,000,000 in 30 years?

The **TVM** (time-value-of-money) **Solver** can be used to answer questions like those posed above.

Reaching Financial Goals

You can use the **TVM Solver** to determine how long to leave money in a savings or investment account to reach a specified financial goal. Chapter 7 provides the steps for using the **TVM Solver**; here are some pointers you may want to use as you follow those steps:

✔ **N** (number of payments):

This is the variable you're solving for. So in Step 4, assign values to all variables except **N**.

✔ **PMT** (amount of payment) and **P/Y** (payments per year):

If you put money into an account and just leave it there without adding to it, then **PMT** is 0. And because you're not making any payments, you can set **P/Y** to 1. This will tell the calculator to evaluate **N** in years.

On the other hand, if you add to the account on a regular basis, then **PMT** is the value of these regular additions to the account and **P/Y** is the number of times a year these additions to the account are made. This is illustrated in the first picture in Figure 8-1. In this picture, the **TVM Solver** is solving the first question posed at the beginning of this chapter.

✔ After you have solved for **N** in Step 8:

If **P/Y** is 1, then the solution for **N** is in years. If **P/Y** isn't 1, then **N** is the total number of payments made. Because **P/Y** payments are made each year, the number of years it takes to reach your financial goal is **N** divided by **P/Y**. Chapter 7 explains how to use a **TVM** value, in particular, how to use these variables in a calculation (as shown in the second picture in Figure 8-1).

Finding **N**   Converting **N** to years

Figure 8-1: Solving for the time needed to reach a financial goal.

Finding Future Value of Money

In the context of a savings or investment account, the future value of your money is the amount of money in the account after a specified time period. The second question posed at the beginning of this chapter is asking for the future value of the account.

To find the future value, follow the steps given in Chapter 7 for using the TVM Solver. Here are some pointers you may want to use:

✔ **FV** (future value):

This is the variable you're solving for. So in Step 4, you have to assign values to all variables except **FV**.

🖊 **PMT** (amount of payment) and **P/Y** (payments per year):

If necessary, read the comments made about **PMT** and **P/Y** in the previous section.

🖊 **N** (number of payments):

N is the number of years you have the account times **P/Y**.

Figure 8-2 illustrates using the **TVM Solver** to answer the second question posed at the beginning of this chapter.

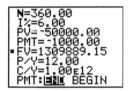

```
N=360.00
I%=6.00
PV=-50000.00
PMT=-1000.00
•FV=1309889.15
P/Y=12.00
C/Y=1.00E12
PMT:END BEGIN
```

Figure 8-2: Finding the future value of money.

Finding Present Value of Money

In the context of a savings or investment account, the *present value* of money is the amount needed to generate a specific amount of money after a specified time period. The third question posed at the beginning of this chapter asks for the present value of an account.

To find the present value, follow the steps given in Chapter 7 for using the TVM Solver. Here are some pointers you may want to use:

🖊 **PV** (present value):

This is the variable you're solving for. Accordingly, in Step 4, assign values to all variables except **PV**.

🖊 **FV** (future value):

This is the amount of money you want to have in the account after the specified time period.

🖊 **PMT** (amount of payment) and **P/Y** (payments per year):

If necessary, read the comments made about **PMT** and **P/Y** in the section titled "Reaching Financial Goals."

🖊 **N** (number of payments):

N is the number of years you have the account times **P/Y**.

Figure 8-3 illustrates using the **TVM Solver** to answer the third question posed at the beginning of this chapter.

```
N=30.00
I%=6.00
•PV=-165298.89
PMT=0.00
FV=1000000.00
P/Y=1.00
C/Y=1.00E12
PMT:BEGIN
```

Figure 8-3: Finding the present value of money.

How many days till Christmas?

To calculate the number of days between two dates:

1. **Press [APPS][1][ALPHA][x⁻¹] to select the dbd command from the Finance application menu. (On the TI-83, press [2nd][x⁻¹][ALPHA][x⁻¹].)**

2. **Enter the first date using either the *mm.ddyy* or the *ddmm.yy* format.**

   In both formats, *mm* is the two-digit number of the month, *dd* is the two-digit number of the day of the month, and *yy* is the last two digits of the year. The date must fall somewhere in the range of years from 1950 through 2049.

   In the U.S.-format *mm.ddyy*, *mm* and *ddyy* are separated by a period; there is no separation between *dd* and *yy*. For example, March 5, 2004, is entered as 03.0504. In the European-format *ddmm.yy*, this date is entered as 0503.04.

3. **Press [.], enter the second date, and then press [)].**

4. **Press [ENTER] to display the number of days between the two dates.**

Using the U.S. date format, the figure given here shows that there are 295 days between March 5, 2004, and Christmas of the same year.

```
dbd(03.0504,12.2
504)
         295.00
```

Part IV
Graphing and Analyzing Functions

The 5th Wave   By Rich Tennant

"WHAT EXACTLY ARE YOU TRYING TO SAY?"

In this part...

*T*his part looks at graphing a function and then analyzing it by tracing the graph or by creating a table of functional values. You get pointers on how to find values associated with the graph, such as minimum and maximum points, points of intersection, and the slope of the curve.

I also show you how to put lines, circles, and text on your graph after you've created it — and save the whole masterpiece for later.

Chapter 9

Graphing Functions

● ●

In This Chapter

▶ Entering functions into the calculator

▶ Graphing functions

▶ Recognizing whether the graph is accurate

▶ Graphing piecewise-defined, and trigonometric functions

▶ Viewing the graph and the function on the same screen

▶ Saving and recalling a graph and its settings in a Graph Database

● ●

*T*he calculator has a variety of features that help you painlessly graph a function. The first step is to enter the function into the calculator. Then to graph the function, you set the viewing window and press GRAPH. Or (better yet) you can use one of several Zoom commands to get the calculator to set the viewing window for you. Finally, after you have graphed the function, you can use Zoom commands to change the look of the graph. For example, you can zoom in or zoom out on a graph the same way that a zoom lens on a camera lets you zoom in or out on the picture you're taking.

Entering Functions

Before you can graph a function, you must enter it into the calculator. The calculator can handle up to ten functions at once, Y_1 through Y_9 and Y_0. To enter functions in the calculator, perform the following steps:

1. **Press MODE and put the calculator in Function mode, as shown in Figure 9-1.**

   To highlight an item in the Mode menu, use the ▶◀▲▼ keys to place the cursor on the item, and then press ENTER. Highlight **Func** in the fourth line to put the calculator in Function mode. (For more about the other items on the Mode menu, refer to Chapter 1.)

Figure 9-1: Setting Function mode.

2. **Press [Y=] to access the Y= editor.**

3. **Enter the definitions of your functions:**

   To erase an entry that appears after Y_n, use the [▶][◀][▲][▼] keys to place the cursor to the right of the equal sign and press [CLEAR]. Then enter your definition for the new function and press [ENTER].

When you're defining functions, the only symbol the calculator allows for the independent variable is the letter X. Press [X,T,θ,n] to enter this letter in the definition of your function. In Figure 9-2, this key was used to enter the functions Y_1, Y_2, Y_4, Y_5, and Y_6.

As a timesaver, when entering functions in the Y= editor, you can reference another function in its definition. (Figure 9-2, for example, shows function Y_3 defined as $-Y_2$.) To paste a function name in the function you're entering in the Y= editor, follow these steps:

1. **Press [VARS][▶] to access the Y-Variables menu.**

2. **Press [1] to access the Function menu.**

3. **Press the number key for the name of the function you want to paste in the definition.**

```
Plot1 Plot2 Plot3
\Y1■(X³-5X²+1)/(
X-5)
\Y2=√(100-X²)
\Y3=-Y2
\Y4=.2X²-8
\Y5=8-.2X²
\Y6=(X-5)²-5
```

Figure 9-2: Examples of entering functions.

Graphing Functions

Here's where your calculator draws pretty pictures. After you have entered the functions into the calculator, as described in the previous section, you can use the following steps to graph the functions:

1. **Turn off any Stat Plots that you don't want to appear in the graph of your functions.**

   The first line in the Y= editor tells you the graphing status of the Stat Plots. (Stat Plots are discussed in Chapter 19.) If **Plot1**, **Plot2**, or **Plot3** is highlighted, then that Stat Plot will be graphed along with the graph of your functions. If it's not highlighted, it won't be graphed. In Figure 9-2, **Plot1** will be graphed along with the functions.

   To turn off a highlighted Stat Plot in the Y= editor, use the ▶◀▲▼ keys to place the cursor on the highlighted Stat Plot and then press ENTER. The same process is used to re-highlight the Stat Plot in order to graph it at a later time.

   When you're graphing functions, Stat Plots can cause problems if they're turned on when you don't really want them to be graphed. The most common symptom of this problem is the ERR: INVALID DIM error message — which by itself gives you almost no insight into what's causing the problem. So if you aren't planning to graph a Stat Plot along with your functions, make sure all Stat Plots are turned off.

2. **Press 2nd ZOOM to access the Format menu.**

3. **Set the format for the graph by using the ▶◀▲▼ keys to place the cursor on the desired format and then press ENTER to highlight it.**

   In the Format menu, each line of the menu will have one item highlighted. An explanation of each menu selection follows:

   • **RectGC and PolarGC:** This gives you a choice between having the coordinates of the location of the cursor displayed in (x, y) rectangular form or in (r, θ) polar form. Select **RectGC** for rectangular form or **PolarGC** for polar form.

   • **CoordOn and CoordOff:** This tells the calculator whether you want to see the coordinates of the cursor location displayed at the bottom of the screen as you move the cursor. Select **CoordOn** if you want to see these coordinates; select **CoordOff** if you don't.

   • **GridOff and GridOn:** If you select **GridOn**, grid points appear in the graph at the intersections of the tick marks on the x- and y-axes (as illustrated in Figure 9-3). If you select **GridOff**, no grid points appear in the graph.

   • **AxesOn and AxesOff:** If you want to see the x- and y-axes on your graph, select **AxesOn**. If you don't want to see them, select **AxesOff**.

• **LabelOff and LabelOn:** If you want the *x*- and *y*-axes to be labeled, select **LabelOn** (as in Figure 9-3). Because the location of the labels isn't ideal, selecting **LabelOff** is usually a wise choice.

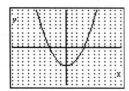

Figure 9-3: A graph with grid points and labeled axes.

• **ExprOn and ExprOff:** If you select **ExprOn**, when you're tracing the graph of a function, the definition of that function appears in the upper left of the screen. If you select **ExprOff** and **CoordOn**, then only the number of the function appears when you trace the function. If you select **ExprOff** and **CoordOff**, then nothing at all appears on the screen to indicate which function you're tracing.

4. Press WINDOW to access the Window editor.

5. After each of the window variables, enter a numerical value that is appropriate for the functions you're graphing. Press ENTER after entering each number.

Figure 9-4 pictures the Window editor when the calculator is in Function mode. The items in this menu determine the viewing window for your graph — in particular, how the *x*- and *y*-axes look on the screen. The following gives an explanation of the variables you must set in this editor:

• **Xmin and Xmax:** These are, respectively, the smallest and largest values of *x* in view on the *x*-axis.

If you don't know what values your graph will need for **Xmin** and **Xmax**, press ZOOM 6 to invoke the **ZStandard** command. This command automatically graphs your functions in the Standard viewing window; the settings for this window appear in Figure 9-4. You can then, if necessary, use the other Zoom commands (described in Chapter 10) to get a better picture of your graph.

• **Xscl:** This is the distance between tick marks on the *x*-axis. (Go easy on the tick marks; using too many makes the axis look like a railroad track. Twenty or fewer tick marks makes for a nice looking axis.)

• **Ymin and Ymax:** These are, respectively, the smallest and largest values of *y* that will be placed on the *y*-axis.

If you have assigned values to **Xmin** and **Xmax** but don't know what values to assign to **Ymin** and **Ymax**, press ZOOM 0 to invoke the **ZoomFit** command. This command uses the **Xmin** and **Xmax** settings to determine the appropriate settings for **Ymin** and **Ymax** and then automatically draws the graph. It does not change the **Yscl** setting. (You must return to the Window editor, if necessary, to adjust this setting yourself.)

- **Yscl:** This is the distance between tick marks on the *y*-axis. (As with the *x*-axis, too many tick marks make the axis look like a railroad track. Fifteen or fewer tick marks is a nice number for the *y*-axis.)

- **Xres:** This setting determines the resolution of the graph. It can be set to any of the integers 1 through 8. When **Xres** is set equal to 1, the calculator evaluates the function at each of the 94 pixels on the *x*-axis and graphs the result. If **Xres** is set equal to 8, the function is evaluated and graphed at every eighth pixel.

Xres is usually set equal to 1. If you're graphing a lot of functions, it may take the calculator a while to graph them at this resolution, but if you change **Xres** to a higher number, you may not get an accurate graph.

If it's taking a long time for the calculator to graph your functions, and this causes you to regret setting **Xres** equal to 1, press ON to terminate the graphing process. You can then go back to the Window editor and adjust the **Xres** setting to a higher number.

6. **Press** GRAPH **to graph the functions.**

```
WINDOW
 Xmin=-10
 Xmax=10
 Xscl=1
 Ymin=-10
 Ymax=10
 Yscl=1
 Xres=1
```

Figure 9-4: Window editor in Function mode.

Graphing Several Functions

If you're graphing several functions at once, it's not easy to determine which graph each function is responsible for. To help clear this up, the calculator allows you to identify the graphs of functions by setting a different graph style for each function. To do this:

1. **Press** Y= **to access the Y= editor.**

2. Use the ►◄▲▼ keys to place the cursor on the icon appearing at the far left of the definition of the function.

3. **Repeatedly press** ENTER **until you get the desired graph style.**

   You have seven styles to choose from: ∖ (Line), ▮ (Thick Line), ▜ (shading above the curve), ▙ (shading below the curve), ⊹ (Path), ⫶ (Animate), and ∴ (Dotted Line). Each time you press ENTER, you get a different style.

   • **Line, Thick Line, and Dotted Line styles:** In Figure 9-2, Y_1 is set to the default line style. Y_2 and Y_3 are set to Thick Line style; and Y_6 is set to Dotted Line style. Figure 9-5 illustrates these styles.

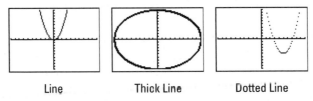

Line   Thick Line   Dotted Line

Figure 9-5: Line style, Thick Line style, and Dotted Line style.

   • **Shading above and below the curve styles:** In Figure 9-2, Y_4 is set to the shading above the curve style and Y_5 is set to the shading below the curve style. The calculator has four shading patterns: vertical lines, horizontal lines, negatively sloping diagonal lines, and positively sloping diagonal lines. You don't get to select the shading pattern. If you're graphing only one function in this style, the calculator uses the vertical line pattern. If you're graphing two functions, the first is graphed in the vertical line pattern and the second in the horizontal line pattern. If you graph three functions in this style, the third appears in the negatively sloping diagonal lines pattern, and so on (as illustrated in Figure 9-6). For custom shading (as pictured in Figure 9-7), see the explanation in Chapter 12.

   Chapter 12 explains how to do custom shading. An example appears in Figure 9-7.

   • **Path and Animated styles:** The Path style, denoted by the ⊹ icon, uses a circle to indicate a point as it's being graphed (as illustrated in Figure 9-8). When the graph is complete, the circle disappears and leaves the graph in Line style.

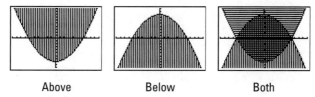

| Above | Below | Both |

Figure 9-6: Shading above the curve, shading below the curve, and two shaded functions.

The animate style, denoted by the ◊ icon, also uses a circle to indicate a point as it's being graphed, but when the graph is complete, no graph appears on the screen. For example, if this style is used, graphing $y = -x^2 + 9$ looks like a movie of the path of a ball thrown in the air.

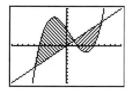

Figure 9-7: Custom shading.

If you don't want the calculator to graph a function in the Y= editor, un-highlight the equal sign in that function. To graph it at a later time, re-highlight the equal sign. This is done in the Y= editor by using the ▶◀▲▼ keys to place the cursor on the equal sign in the definition of the function and then pressing ENTER to toggle the equal sign between highlighted and un-highlighted. In the example in Figure 9-2, the calculator won't graph functions $\mathbf{Y_2}$ through $\mathbf{Y_6}$.

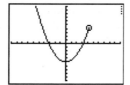

Figure 9-8: Path style.

Is Your Graph Accurate?

The calculator can do only what you tell it to do — which doesn't always produce an accurate graph. The three main causes of inaccurate graphs, and their solutions, are the following:

✔ **The graph is distorted by the size of the screen.**

Because the calculator screen isn't square, circles don't look like circles unless the viewing window is properly set. How do you properly set the viewing window? No problem! Just graph the function as described earlier in this chapter, and then press ZOOM 5 to invoke the **ZSquare** command. **ZSquare** readjusts the window settings for you and re-graphs the function in a viewing window in which circles look like circles. Figure 9-9 illustrates this. (The circle being drawn in each of these figures is the circle defined by Y_2 and Y_3, as shown in Figure 9-2.)

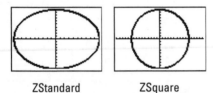

ZStandard   ZSquare

Figure 9-9: A circle graphed using ZStandard and then using ZSquare.

✔ **The viewing window is too small or too big.**

If you don't know what the graph should look like, then after graphing it you should zoom out to see more of the graph or zoom in to see a smaller portion of the graph. To do this, press ZOOM 3 to zoom out, or press ZOOM 2 to zoom in.

Then use the ▶ ◀ ▲ ▼ keys to move the cursor to the point from which you want to zoom out or in, and press ENTER. It's just like using a camera. The point you want to move the cursor to is the focal point.

After zooming in or out, you may have to adjust the window settings, as described earlier in this chapter.

As an example, Figure 9-10 shows the progression of graphing Y_1 in Figure 9-2. It was first graphed in the Standard viewing window. Then it was zoomed out from the point (0, 10). And finally, the window settings were adjusted to get a better picture of the graph.

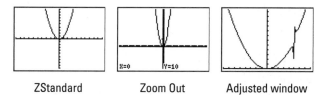

ZStandard   Zoom Out   Adjusted window

Figure 9-10: A graph using ZStandard, then Zoom Out, and then an adjusted viewing window.

✔ **Vertical asymptotes may not be recognizable.**

In all graph styles except Animate and Dotted Line, the calculator graphs one point, and then the next point, and connects those two points with a line segment. This sometimes causes vertical asymptotes to appear on the graph. The last graph in Figure 9-10 illustrates an example of when a vertical asymptote is present. Don't mistake this almost-vertical line for a part of the graph. It's not; it's just a vertical asymptote.

In a different viewing window, the vertical asymptote may not even appear. This happens when the calculator graphs one point, but the next point is undefined because the *x* value of that point is exactly at the location of the vertical asymptote. Figure 9-11 gives an example of re-graphing the last graph shown in Figure 9-10, using a viewing window in which the vertical asymptote does not appear.

If you want to ensure that vertical asymptotes don't appear on your graph, graph the function in the Dotted Line style described in the previous section. For an explanation of how you can draw a vertical asymptote on a graph, see Chapter 12.

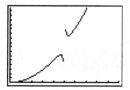

Figure 9-11: A viewing window in which no vertical asymptotes appear.

Piecewise-Defined Functions

When graphing a piecewise-defined function, some people cop out and simply graph each of the separate functions that appear in it. But this does not result in an accurate graph. To accurately graph the following piecewise-defined function, perform the following steps:

$$y = \begin{cases} y_1, \ x < a \\ y_2, \ a \le x \le b \\ y_3, \ x > b \end{cases}$$

1. **Enter the functions Y$_1$, Y$_2$, and Y$_3$ in the Y= editor.**

   Entering functions in the Y= editor is explained earlier in the chapter. Examples of three such functions appear in Figure 9-12.

2. **Turn off Y$_1$, Y$_2$, and Y$_3$ by un-highlighting their equal signs.**

   This is done in the Y= editor by using the ▶◀▲▼ keys to place the cursor on the equal sign in the definition of the function and then pressing ENTER to toggle the equal sign between highlighted and un-highlighted. The calculator graphs a function only when its equal sign is highlighted. An example of this appears in Figure 9-12.

3. **Enter the piecewise-defined function in Y$_4$ as (Y$_1$)(x < a) + (Y$_2$)(a ≤ x)(x ≤ b) + (Y$_3$)(x > b).**

   The function must be entered as it appears above, complete with parentheses. **(Y$_1$)(x < a)** tells the calculator to graph the function Y$_1$ for $x < a$ and **(Y$_2$)(a ≤ x)(x ≤ b)** tells it to graph Y$_2$ for $a \le x \le$ b.

   • **To enter Y$_1$, Y$_2$, and Y$_3$:** Press VARS ▶ 1 and then press the number of the function you want to use in the definition.

   • **To enter the inequalities:** Press 2nd MATH to access the Test menu, and then press the number of the inequality you want to use in the definition. For example, to enter the less-than (<) symbol, press 2nd MATH 5.

   An example of entering this function appears in Figure 9-12.

4. **Press ENTER.**

5. **Graph the piecewise-defined function.**

   An earlier section of this chapter explains how to graph functions. An example graph of a piecewise-defined function appears in Figure 9-12.

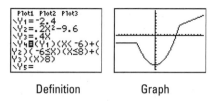

| Definition | Graph |

Figure 9-12: Defining and graphing a piecewise-defined function.

Graphing Trig Functions

The calculator has built-in features especially designed for graphing trigonometric functions. They produce graphs that look like graphs you see in text books; and when you trace these graphs, the *x*-coordinate of the tracing point is always given as a fractional multiple of π. To use these features when graphing trigonometric functions, follow these steps:

1. **Put the calculator in Function and Radian mode.**

   Press MODE. In the third line, highlight **Radian**, and in the fourth highlight **Func**. (To highlight an item in the Mode menu, use the ►◄▲▼ keys to place the cursor on the item, and then press ENTER.)

2. **Enter your trigonometric functions into the Y= editor.**

   Entering functions in the Y= editor is explained earlier in the chapter.

3. **Press ZOOM 7 to graph the function.**

   ZOOM 7 invokes the **ZTrig** command which graphs the function in a viewing window in which $-47\pi/24 \le x \le 47\pi/24$ and $-4 \le y \le 4$. It also sets the tick marks on the *x*-axis to multiples of π/2.

   When you trace a function graphed in a **ZTrig** window, the *x*-coordinate of the trace cursor will be a multiple of π/24. (Tracing is explained in the next chapter.)

Viewing the Function and Graph on the Same Screen

If you're planning to play around with the definition of a function you're graphing, it's quite handy to have both the Y= editor and the graph on the same screen. That way you can edit the definition of your function and see the effect your editing has on your graph. To do so, follow these steps:

1. **Put the calculator in Horizontal mode.**

   Press MODE and highlight **Horiz** in the last line of the menu, as illustrated in Figure 9-13. To highlight an item in the Mode menu, use the ▶◀▲▼ keys to place the cursor on the item, and then press ENTER.

2. **Press Y=.**

   The Graph window appears at the top of the screen and the Y= editor at the bottom of the screen.

3. **Enter or edit a function in the Y= editor.**

   Entering functions in the Y= editor is explained earlier in this chapter. Editing expressions is explained in Chapter 1.

4. **Graph the function.**

   Graphing functions is explained earlier in this chapter.

To edit or enter a function, press Y=. To see the resulting graph, press GRAPH.

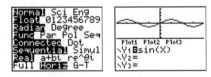

Mode   Split screen

Figure 9-13: Function and graph on the same screen.

Saving and Recalling a Graph

If you want to save just a picture of your graph, see Chapter 12. However, remember that pictures of graphs are not interactive. You cannot, for example, trace them or resize the viewing window. All you can do is look at them.

On the other hand, if you want your recalled graph to be interactive, save it as a Graph Database. This way the calculator saves the Graph Mode, Window, Format, and Y= editor settings. It does not, however, save the split-screen settings (**Horiz** and **G-T**) entered in the last line of the Mode menu. Here's how to save, delete, and recall a graph in a Graph Database:

To save a Graph Database, perform the following steps:

1. **Press** [2nd][PRGM][▶][▶] **to access the Draw Store menu.**

2. **Press** [3] **to store your graph as a Graph Database.**

3. **Enter an integer 0 through 9.**

   The calculator can store up to 10 Graph Databases. If, for example, you enter the number **5**, your Graph Database is stored in the calculator as **GDB5**.

   If you save your Graph Database as **GDB5** without realizing that you had previously stored another Graph Database as **GDB5**, the calculator — without warning or asking your permission — erases the old **GDB5** and replaces it with the new **GDB5**. To see a list of the Graph Databases already stored in your calculator, press [2nd][+][2][9].

   If you already have ten Graph Databases stored in your calculator and don't want to sacrifice any of them, consider saving some of them on your PC. Chapter 22 describes how to do this.

4. **Press** [ENTER].

To delete a Graph Database from your calculator, perform the following steps:

1. **Press** [2nd][+] **to access the Memory menu.**

2. **Press** [2] **to access the Mem Mgt/Del menu.**

3. **Press** [9] **to access the GBD files stored in the calculator.**

4. **If necessary, repeatedly press** [▼] **to move the indicator to the GBD you want to delete.**

5. **Press** [DEL].

   If there is more than one Graph Database stored in your calculator, you are asked whether or not you really want to delete this item. Press [2] if you want it deleted, or press [1] if you have changed your mind about deleting it.

6. **Press** [2nd][MODE] **to exit this menu and return to the home screen.**

To recall a saved Graph Database, perform the following steps:

1. **Press** [2nd][PRGM][▶][▶] **to access the Draw Store menu.**

2. **Press** [4] **to recall your Graph Database.**

3. Enter the number of your stored Graph Database.

4. Press ENTER.

When you recall a Graph Database, the Mode, Window, Format, and Y= editor settings in your calculator change to those saved in the Graph Database. If you don't want to lose the settings you have in the calculator, save them in another Graph Database before recalling your saved Graph Database. (Saving a Graph Database is described earlier in this section.)

Chapter 10

Exploring Functions

. . . .

In This Chapter

▶ Using Zoom commands

▶ Tracing the graph of a function

▶ Constructing tables of functional values

▶ Creating and clearing user-defined tables

▶ Viewing graphs and tables on the same screen

. . . .

*T*he calculator has three very useful features that help you explore the graph of a function: zooming, tracing, and creating tables of functional values. Zooming allows you to quickly adjust the viewing window for the graph so that you can get a better idea of the nature of the graph. Tracing shows you the coordinates of the points that make up the graph. And creating a table — well, I'm sure you already know what that shows you. The following sections explain how to use each of these features.

Using Zoom Commands

After you've graphed your functions (as described in Chapter 9), you can use Zoom commands to adjust the view of your graph. Press ZOOM to see the ten Zoom commands that you can use. The following list explains the Zoom commands and how to use them:

  ✔ **Zoom commands that help you to initially graph or regraph your function in a preset viewing window:**

  • **ZStandard:** This command graphs your function in a preset viewing window where $-10 \leq x \leq 10$ and $-10 \leq y \leq 10$. You access it by pressing ZOOM 6.

  This is a nice Zoom command to use when you haven't the slightest idea what size viewing window to use for your function. After graphing the function using **ZStandard**, you can, if necessary, use the **Zoom In** and

Zoom Out commands to get a better idea of the nature of the graph. Using **Zoom In** and **Zoom Out** are described later in this section.

- **ZDecimal:** This command graphs your function in a preset viewing window where $-4.7 \le x \le 4.7$ and $-3.1 \le y \le 3.1$. It is accessed by pressing ZOOM 4.

    When you trace a function graphed in a **ZDecimal** window, the x-coordinate of the trace cursor will be a multiple of 0.1. Tracing is explained in the next section.

- **ZTrig:** This command, which is most useful when graphing trigonometric functions, graphs your function in a preset viewing window where $-47\pi/24 \le x \le 47\pi/24$ and $-4 \le y \le 4$. It also sets the tick marks on the x-axis to multiples of $\pi/2$. You access **ZTrig** by pressing ZOOM 7.

    When you trace a function graphed in a **ZTrig** window, the x-coordinate of the trace cursor will be a multiple of $\pi/24$. Tracing is explained in the next section.

To use the zoom commands described above, enter your function into the calculator (as described in Chapter 9), press ZOOM, and then press the key for the number of the command. The graph automatically appears.

✔ **Zoom commands that help you find an appropriate viewing window for the graph of your functions:**

- **ZoomFit:** This is my favorite Zoom command. If you know how you want to set the x-axis, **ZoomFit** automatically figures out the appropriate settings for the y-axis.

    To use **ZoomFit**, press WINDOW and enter the values you want for **Xmin**, **Xmax**, and **Xscl**. Then press ZOOM 0 to get **ZoomFit** to figure out the y-settings and graph your function. **ZoomFit** does not figure out an appropriate setting for **Yscl**, so you may want to go back to the Window editor and adjust this value. The Window editor is discussed in Chapter 9.

- **ZoomStat:** If you're graphing functions, this command is useless. But if you're graphing Stat Plots (as explained in Chapter 19), this command finds the appropriate viewing window for your plots. See Chapter 18 for information on how this works.

✔ **Zoom commands that readjust the viewing window of an already-graphed function:**

- **ZSquare:** Because the calculator screen isn't perfectly square, graphed circles won't look like real circles unless the viewing window is properly set. **ZSquare** readjusts the existing Window settings for you and then regraphs the function in a viewing window in which circles look like circles.

To use **ZSquare**, graph the function as described in Chapter 9, and then press ZOOM⑤. The graph automatically appears.

- **ZInteger:** This command is quite useful when you want the trace cursor to trace your functions using integer values of the x-coordinate, such as when graphing a function that defines a sequence. (Tracing is explained in the next section.) **ZInteger** readjusts the existing Window settings and regraphs the function in a viewing window in which the trace cursor displays integer values for the x-coordinate.

   To use **ZInteger**, graph the function as described in Chapter 9, and then press ZOOM⑧. Use the ▶◀▲▼ keys to move the cursor to the spot on the screen that will become the center of the new screen. Then press ENTER. The graph is redrawn centered at the cursor location.

✔ **Zoom commands that zoom in or zoom out from an already graphed function.**

- **Zoom In and Zoom Out:** After the graph is drawn (as described in the previous chapter), these commands allow you to zoom in on a portion of the graph or to zoom out from the graph. They work very much like a zoom lens on a camera.

   Press ZOOM② to zoom in or press ZOOM③ to zoom out. Then use the ▶◀▲▼ keys to move the cursor to the spot on the screen from which you want to zoom in or zoom out. Then press ENTER. The graph is redrawn centered at the cursor location.

   You can press ENTER again to zoom in closer or to zoom out one more time. Press CLEAR when you're finished zooming in or zooming out. You may have to adjust the window settings, as described earlier in the Chapter 9.

- **ZBox:** This command allows you to define a new viewing window for a portion of your graph by enclosing it in a box, as illustrated in Figure 10-1. The box becomes the new viewing window.

   To construct the box, press ZOOM① and use the ▶◀▲▼ keys to move the cursor to the spot where you want one corner of the box to be located. Press ENTER to anchor that corner of the box. Then use the ▶◀▲▼

keys to construct the rest of the box. When you press these keys, the calculator draws the sides of the box. Press ⌈ENTER⌉ when you're finished drawing the box. The graph is then redrawn in the window defined by your box (as shown in Figure 10-1).

When you use **ZBox**, if you don't like the size of the box you get, you can use any of the ⌈▶⌉⌈◀⌉⌈▲⌉⌈▼⌉ keys to resize the box. If you don't like the location of the corner you anchored, press ⌈CLEAR⌉ and start over.

When you use **ZBox**, ⌈ENTER⌉ is pressed only two times. The first time you press it's to anchor a corner of the zoom box. The next time you press ⌈ENTER⌉ is when you're finished drawing the box, and are ready to have the calculator redraw the graph.

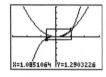 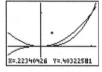

The box   Redrawn graph

Figure 10-1: Constructing the Zoom Box and redrawing the graph.

Tracing a Graph

After you have graphed your function, as described in the previous chapter, you can press ⌈TRACE⌉ and then use ⌈▶⌉ and ⌈◀⌉ to more closely investigate the function.

If you use only the ⌈▶⌉⌈◀⌉⌈▲⌉⌈▼⌉ keys instead of ⌈TRACE⌉ to locate a point on a graph, all you will get is an *approximation* of the location of that point. You rarely get an actual point on the graph. So always use ⌈TRACE⌉ to identify points on a graph.

The following list describes what you see, or don't see, as you trace a graph:

✔ **The definition of the function:** The function you're tracing is displayed at the top of the screen, provided the calculator is in **ExprOn** format, as discussed in Chapter 9. If the Format menu is set to **ExprOff** and **CoordOn**, then the Y= editor number of the function appears at the top right of the screen. If the Format menu is set to **ExprOff** and **CoordOff**, then tracing the graph is useless because all you see is a cursor moving on the graph. The calculator won't tell you which function you're tracing, nor will it tell you the coordinates of the cursor location. (The Format menu and Y= editor are described in Chapter 9.)

If you've graphed more than one function and you would like to trace a different function, press ▲. Each time you press this key, the cursor jumps to another function. Eventually it jumps back to the original function.

✔ **The values of *x* and *y*:** At the bottom of the screen you see the values of the *x*- and *y*- coordinates that define the cursor location, provided the calculator is in **CoordOn** format (as discussed in Chapter 9). In the **PolarGC** format, the coordinates of this point display in polar form.

When you press [TRACE], the cursor is placed on the graph at the point having an *x*-coordinate that is approximately midway between **Xmin** and **Xmax**. If the *y*-coordinate of the cursor location isn't between **Ymin** and **Ymax**, then the cursor does not appear on the screen. The sidebar "Panning in Function mode" tells you how to correct this situation.

Each time you press ▶, the cursor moves right to the next plotted point on the graph, and the coordinates of that point are displayed at the bottom of the screen. If you press ◀, the cursor moves left to the previously plotted point. And if you press ▲ to trace a different function, the tracing of that function starts at the point on the graph that has the *x*-coordinate displayed on-screen before you pressed this key.

Press [CLEAR] to terminate tracing the graph. This also removes the name of the function and the coordinates of the cursor from the screen.

Un-zooming

If you used a Zoom command to redraw a graph and then want to undo what that command did to the graph, then:

1. Press [ZOOM]▶ **to access the Zoom Memory menu.**

2. Press [1] **to select ZPrevious.**

The graph is redrawn as it appeared in the previous viewing window.

Panning in Function mode

When you're tracing a function and the cursor hits the top or bottom of the screen, you will still see the coordinates of the cursor location displayed at the bottom of the screen but you won't see the cursor itself on the screen because the viewing window is too small. Press [ENTER] to get the calculator to adjust the viewing window to a viewing window that is centered about the cursor location. If the function you were tracing isn't displayed at the top of the screen, repeatedly press [▲] until it is. The trace cursor then appears in the middle of the screen and you can use [▶] and [◀] to continue tracing the graph.

When you're tracing a function and the cursor hits the left or right side of the screen, the calculator automatically pans left or right. It also appropriately adjusts the values assigned to **Xmin** and **Xmax** in the Window editor, but it does not change the values of **Ymin** and **Ymax**. So you may not see the cursor on the screen. If this happens, follow the directions in the previous paragraph to see both the function and the cursor on the screen.

When you're using TRACE, if you want to start tracing your function at a specific value of the independent variable x, just key in that value and press [ENTER] when you're finished. (The value you assign to x must be between **Xmin** and **Xmax**; if it's not, you get an error message.) After you press [ENTER], the trace cursor moves to the point on the graph having the x-coordinate you just entered. If that point isn't on the portion of the graph appearing on the screen, the sidebar "Panning in Function mode" tells you how to get the cursor and the graph in the same viewing window.

If the name of the function and the values of x and y are interfering with your view of the graph when you use TRACE, increase the height of the screen by pressing [WINDOW] and then decrease the value of **Ymin** and increase the value of **Ymax**.

Displaying Functions in a Table

After you've entered the functions in the Y= editor (as described in the previous chapter), you can have the calculator create a table of functional values. To create a table, perform the following steps:

1. **Highlight the equal sign of those functions in the Y= editor that you want to appear in the table.**

   Only those functions in the Y= editor that are defined with a highlighted equal sign appear in the table. To highlight or un-highlight an equal sign, press Y=, use the ▶◀▲▼ keys to place the cursor on the equal sign in the definition of the function, and then press ENTER to toggle the equal sign between highlighted and un-highlighted.

2. **Press 2nd WINDOW to access the Table Setup editor (shown in Figure 10-2).**

```
TABLE SETUP
 TblStart=5
 ΔTbl=-1
Indpnt: Auto Ask
Depend: Auto Ask
```

Figure 10-2: The Table Setup editor.

3. **Enter a number in TblStart and then press ENTER.**

TblStart is the first value of the independent variable x to appear in the table. In Figure 10-2, **TblStart** is assigned the value 5.

To enter the number you have chosen for **TblStart**, place the cursor on the number appearing after the equal sign, press the number keys to enter your new number, and then press ENTER.

4. **Enter a number in ΔTbl and then press ENTER.**

ΔTbl gives the increment for the independent variable x. In Figure 10-2, **ΔTbl** is assigned the value –1.

To enter the number you have chosen for **ΔTbl**, place the cursor on the number appearing after the equal sign, press the number keys to enter your new number, and then press ENTER.

5. **Set the mode for Indpnt and Depend.**

To change the mode of either **Indpnt** or **Depend**, use the ▶◀▲▼ keys to place the cursor on the desired mode, either **Auto** or **Ask**, and then press ENTER.

To have the calculator automatically generate the table for you, put both **Indpnt** and **Depend** in **Auto** mode. The table in Figure 10-3 was constructed in this fashion.

If you want to create a user-defined table in which you specify which values of the independent variable x appear in the table — and then have the calculator figure out the corresponding values of the functions — put **Indpnt** in **Ask** mode and **Depend** in **Auto** mode. (How you construct the table is explained in Step 6.) The table in Figure 10-4 was constructed in this fashion.

X	Y₁	Y₂
5	ERROR	8.6603
4	15	9.1652
3	8.5	9.5394
2	3.6667	9.798
1	.75	9.9499
0	-.2	10
-1	.83333	9.9499

X=5

Figure 10-3: A table automatically generated by the calculator.

For a user-defined table, you don't have to assign values to **TblStart** and **ΔTbl** in the Table Setup editor.

The other combinations of mode settings for **Indpnt** and **Depend** are not all that useful unless you want to play a quick round of "Guess the *y*-coordinate."

X	Y₁	Y₂
5	ERROR	8.6603
4.5	18.25	8.9303
5.5	32.25	8.3518
-7	48.917	7.1414

X=

Figure 10-4: A user-generated table.

6. **Press** 2nd GRAPH **to display the table.**

When you display the table, what you see on the screen depends on the modes you set for **Indpnt** and **Depend** in Step 5. And what you can do with the table also depends on these modes. Here's what you see and what you can do:

• **An automatically generated table:**

If **Indpnt** and **Depend** are both in **Auto** mode, then when you press 2nd GRAPH, the table is automatically generated. To display rows in the table beyond the last row on the screen, repeatedly press ▼ until they appear. You can repeatedly press ▲ to display rows above the first row on the screen.

Each time the calculator redisplays a table with a different set of rows, it also automatically resets **TblStart** to the value of *x* that appears in the first row of the newly displayed table. To return the table to its original state, press 2nd WINDOW to access the Table Setup editor, and then change the value that the calculator assigned to **TblStart**.

If you're constructing a table for more than two functions, only the first two functions appear on the screen. To see the other functions, repeatedly press ▶ until they appear. This causes one or more of the initial functions to disappear. To see them again, repeatedly press ◀ until they appear.

- **A user-generated table:**

   If you put **Indpnt** in **Ask** mode and **Depend** in **Auto** mode so that you can generate your own table; then when you display the table, it should be empty. If it's not, clear the table (as described in the next section).

   In an empty table, key in the first value of the independent variable x that you want to appear in the table, and then press ENTER. The corresponding y-values of the functions in the table automatically appear. Key in the next value of x you want in the table and press ENTER, and so on. The values of x that you place in the first column of the table don't have to be in any specific order, nor do they have to be between the **Xmin** and **Xmax** settings in the Window editor.

The word ERROR appearing in a table doesn't indicate that the creator of the table has done something wrong. It indicates that either the function is undefined or the corresponding value of x is not a real-valued number. This is illustrated in Figures 10-3 and 10-4.

While displaying the table of functional values, you can edit the definition of a function without going back to the Y= editor. To do this, use the ▶◀▲▼ keys to place the cursor on the column heading for that function and then press ENTER. Edit the definition of the function (editing expressions is explained in Chapter 1) and press ENTER when you're finished. The calculator automatically updates the table and the definition of the function in the Y= editor.

Clearing a Table

Not all tables are created alike. An automatically generated table, for example, cannot be cleared. To change the contents of such a table, you have to change the values assigned to **TblStart** and **ΔTbl** in the Table Setup editor. After you have created a user-defined table, however, you can perform the following steps to clear its contents:

1. **Press** 2nd WINDOW **to access the Table Setup editor and then set Indpend to Auto.**

2. **Press** 2nd GRAPH **to display an automatically generated table.**

3. **Press** 2nd WINDOW **and set Indpend back to Ask.**

4. **Press** 2nd GRAPH **to display an empty table.**

Viewing the Table and the Graph on the Same Screen

After you have graphed your functions, and created a table of functional values, you can view the graph and the table on the same screen. To do so, follow these steps:

1. **Press** MODE.

2. **Put the calculator in G-T screen mode.**

   To do so, use the ► ◄ ▲ ▼ keys to place the cursor on **G-T** in the lower-right corner of the Mode menu, and then press ENTER to highlight it. This is illustrated in Figure 10-5.

Figure 10-5: Setting the mode for viewing a graph and a table.

3. **Press** GRAPH.

   After you press GRAPH, the graph and the table appear on the same screen (as shown in Figure 10-6).

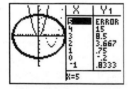

Figure 10-6: A Graph-Table split screen.

If you press any key used in graphing functions, such as ZOOM or TRACE, the cursor becomes active on the graph side of the screen. To return the cursor to the table, press 2nd GRAPH.

If you press $\boxed{\text{TRACE}}$ and then use the $\boxed{\blacktriangleright}\boxed{\blacktriangleleft}\boxed{\blacktriangle}\boxed{\blacktriangledown}$ keys to trace the graph, the value of the independent variable x corresponding to the cursor location on the graph is highlighted in the table and the column for the function you're tracing appears next to it. If necessary, the calculator updates the table so you can see that row in the table (as in Figure 10-7).

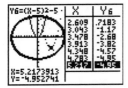

Figure 10-7: Using TRACE in Graph-Table mode.

To view the graph or the table in full screen mode, you can use these steps:

1. **Press** $\boxed{\text{MODE}}$.

2. **Put the calculator in Full screen mode.**

   To do so, use the $\boxed{\blacktriangleright}\boxed{\blacktriangleleft}\boxed{\blacktriangle}\boxed{\blacktriangledown}$ keys to place the cursor on **Full** in the bottom left hand corner of the Mode menu and press $\boxed{\text{ENTER}}$ to highlight it.

3. **Press** $\boxed{\text{GRAPH}}$ **to see the graph, or press** $\boxed{\text{2nd}}\boxed{\text{GRAPH}}$ **to see the table.**

Chapter 11

Evaluating Functions

· ·

In This Chapter

▶ Finding the value of a function

▶ Finding the zeros (*x*-intercepts) of a function

▶ Finding the maximum and minimum values of a function

▶ Finding the point of intersection of two functions

▶ Finding the slope of a tangent to the graph of a function

▶ Finding the value of the definite integral of a function

· ·

*A*fter graphing a function (as described in Chapter 9), you can use the options on the Calculate menu to find the value of the function at a specified value of *x*, to find the zeros (*x*-intercepts) of the function, and to find the maximum and minimum values of the function. You can even find the derivative of the function at a specified value of *x*, or you can evaluate a definite integral of the function. This, in turn, enables you to find the slope of the tangent to the graph of the function at a specified value of *x* or to find the area between the graph and the *x*-axis. Moreover, if you have graphed two functions, there is an option on the Calculate menu that finds the coordinates of these two functions' points of intersection.

The rest of this chapter tells you how to use the Calculate menu to find these values. But be warned: The calculator is not perfect. In most cases, using the options on the Calculate menu yields only an approximation of the true value (albeit a very *good* approximation).

Finding the Value of a Function

When you trace the graph of a function, the trace cursor doesn't hit every point on the graph. So tracing is not a reliable way of finding the value of a function at a specified value of the independent variable *x*. The **CALC** menu, however, contains a command that will evaluate a function at any specified *x*-value. To access and use this command, perform the following steps:

1. **Graph the functions in a viewing window that contains the specified value of x.**

   Graphing functions and setting the viewing window are explained in Chapter 9. To get a viewing window containing the specified value of x, that value must be between **Xmin** and **Xmax**.

2. **Set the Format menu to ExprOn and CoordOn.**

   Setting the Format menu is explained in Chapter 9.

3. **Press** [2nd][TRACE] **to access the Calculate menu.**

4. **Press** [1] **to select the value option.**

5. **Enter the specified value of x.**

   When using the **value** command to evaluate a function at a specified value of x, that value must be an x-value that appears on the x-axis of the displayed graph — that is, it must be between **Xmin** and **Xmax**. If it isn't, you get an error message.

   Use the keypad to enter the value of x (as illustrated in the first graph in Figure 11-1). If you make a mistake when entering your number, press [CLEAR] and re-enter the number.

6. **Press** [ENTER].

   After you press [ENTER], the first highlighted function in the Y= editor appears at the top of the screen, the cursor appears on the graph of that function at the specified value of x, and the coordinates of the cursor appear at the bottom of the screen. This is illustrated in the second graph in Figure 11-1.

7. **Repeatedly press** [▲] **to see the value of the other graphed functions at your specified value of x.**

   Each time you press [▲], the name of the function being evaluated appears at the top of the screen and the coordinates of the cursor location appears at the bottom of the screen. This is illustrated in the third graph in Figure 11-1.

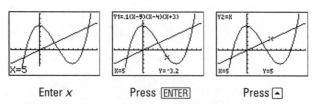

Figure 11-1: Steps in evaluating two functions at a specified value of x.

After using the **value** command to evaluate your functions at one value of x, you can evaluate your functions at another value of x by keying in the new value and then pressing ▲. Pressing any function key (such as ▲ or TRACE) *after* evaluating a function deactivates the **value** command.

TIP

If you plan to evaluate functions at several specified values of x, consider constructing a user-defined table of functional values (as explained in Chapter 10).

Finding the Zeros of a Function

The *zeros* of the function $y = f(x)$ are the solutions to the equation $f(x) = 0$. Because $y = 0$ at these solutions, these zeros (solutions) are really just the x-coordinates of the x-intercepts of the graph of $y = f(x)$. (An x-intercept is a point where the graph crosses or touches the x-axis.) To find a zero of a function, perform the following steps:

1. **Graph the function in a viewing window that contains the zeros of the function.**

   Graphing a function and finding an appropriate viewing window are explained in Chapter 9. To get a viewing window containing a zero of the function, that zero must be between **Xmin** and **Xmax** and the x-intercept at that zero must be visible on the graph.

2. **Set the Format menu to ExprOn and CoordOn.**

   Setting the Format menu is explained in Chapter 9.

3. **Press** 2nd TRACE **to access the Calculate menu.**

4. **Press** 2 **to select the zero option.**

5. **If necessary, repeatedly press** ▲ **until the appropriate function appears at the top of the screen.**

6. **Set the Left Bound for the zero you desire to find.**

   To do so, use the ◄ and ► keys to place the cursor on the graph a little to the left of the zero, and then press ENTER. A Left Bound indicator appears at the top of the screen (as illustrated by the triangle in the first graph of Figure 11-2).

7. **Set the Right Bound for the zero.**

   To do so, use the ◄ and ► keys to place the cursor on the graph a little to the right of the zero, and then press ENTER. A Right Bound indicator appears at the top of the screen, as shown by the right triangle in the second graph of Figure 11-2.

8. **Tell the calculator where you guess the zero is located.**

This guess is necessary because the calculator uses a numerical routine for finding a zero. The routine is an iterative process that requires a seed (guess) to get it started. The closer the seed is to the zero, the faster the routine finds the zero. To do this, use the ◀ and ▶ keys to place the cursor on the graph as close to the zero as possible, and then press ENTER. The value of the zero appears at the bottom of the screen, as shown in the third graph of Figure 11-2.

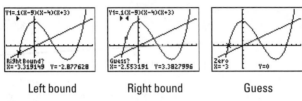

Left bound   Right bound   Guess

Figure 11-2: Steps in finding the zero of a function.

Finding Min & Max

Finding the maximum or minimum point on a graph has many useful applications. For example, the maximum point on the graph of a profit function not only tells you the maximum profit (the y-coordinate), it also tells you how many items (the x-coordinate) the company must manufacture to achieve this profit. To find the minimum or maximum value of a function, perform the following steps:

1. **Graph the function in a viewing window that contains the minimum and/or maximum values of the function.**

   Graphing a function and finding an appropriate viewing window are explained in Chapter 9.

2. **Set the Format menu to ExprOn and CoordOn.**

   Setting the Format menu is explained in Chapter 9.

3. **Press 2nd TRACE to access the Calculate menu.**

4. **Press 3 to find the minimum, or press 4 to find the maximum.**

5. **If necessary, repeatedly press ▲ until the appropriate function appears at the top of the screen.**

6. **Set the Left Bound of the minimum or maximum point.**

   To do so, use the ◀ and ▶ keys to place the cursor on the graph a little to the left of the location of the minimum or maximum point, and then press ENTER. A *left bound indica-tor* (the triangle shown in the first graph of Figure 11-3) appears at the top of the screen.

7. Set the Right Bound for the zero.

To do so, use the ◀ and ▶ keys to place the cursor on the graph a little to the right of the location of the minimum or maximum point, and then press ENTER. A *right bound indicator* (the rightmost triangle in the second graph of Figure 11-3) appears at the top of the screen.

8. Tell the calculator where you guess the min or max is located.

To do so, use the ◀ and ▶ keys to place the cursor on the graph as close to the location of the minimum or maximum point as possible, and then press ENTER. The coordinates of the minimum or maximum point appears at the bottom of the screen (as in the third graph of Figure 11-3).

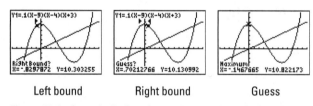

| Left bound | Right bound | Guess |

Figure 11-3: Steps in finding the maximum value of a function.

Finding Points of Intersection

Using the ▶◀▲▼ keys to locate the point of intersection of two graphs gives you an *approximation* of that point, but this method rarely gives you the actual point of intersection. To accurately find the coordinates of the point where two functions intersect, per-form the following steps:

1. Graph the functions in a viewing window that contains the point of intersection of the functions.

Graphing a function and finding an appropriate viewing window are explained in Chapter 9.

2. Set the Format menu to ExprOn and CoordOn.

Setting the Format menu is explained in Chapter 9.

3. Press 2nd TRACE to access the Calculate menu.

4. Press 5 to select the intersect option.

5. Select the first function.

If the name of one of the intersecting functions does not appear at the top of the screen, repeatedly press ⬆ until it does. This is illustrated in the first graph of Figure 11-4. When the cursor is on one of the intersecting functions, press ENTER to select it.

6. Select the second function.

If the calculator does not automatically display the name of the second intersecting function at the top of the screen, repeatedly press ⬆ until it does. This is illustrated in the second graph of Figure 11-4. When the cursor is on the second intersecting function, press ENTER to select it.

7. Use the ◄ and ► to move the cursor as close to the point of intersection as possible.

This is illustrated in the third graph in Figure 11-4.

8. Press ENTER to display the coordinates of the point of intersection.

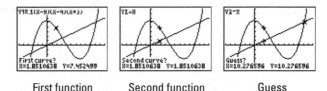

First function   Second function   Guess

Figure 11-4: Steps in finding a point of intersection.

Finding the Slope of a Curve

The calculator is not equipped to find the derivative of a function. For example, it cannot tell you that the derivative of x^2 is $2x$. But it is equipped with a numerical routine that evaluates the derivative at a specified value of x. This numerical value of the derivative is the slope of the tangent to the graph of the function at the specified x-value. It is also called the slope of the curve. To find the slope (derivative) of a function at a specified value of x, perform the following steps:

1. Graph the function in a viewing window that contains the specified value of x.

Graphing a function and setting the viewing window are explained in Chapter 9. To get a viewing window containing the specified value of x, that value must be between **Xmin** and **Xmax**.

2. **Set the Format menu to ExprOn and CoordOn.**

   Setting the Format menu is explained in Chapter 9.

3. **Press 2nd TRACE to access the Calculate menu.**

4. **Press 6 to select the dy/dx option.**

5. **If necessary, repeatedly press ▲ until the appropriate function appears at the top of the screen.**

   This is illustrated in the first graph in Figure 11-5.

6. **Enter the specified value of x.**

   To do so, use the keypad to enter the value of x. As you use the keypad, **X=** appears, replacing the coordinates of the cursor location appearing at the bottom of the screen in the previous step. The number you key in appears after **X=**. This is illustrated in the second graph in Figure 11-5. If you make a mistake when entering your number, press CLEAR and re-enter the number.

   If you are interested only in finding the slope of the function in a general area of the function instead of at a specific value of x, and then instead of entering a value of x, just use the ◄ and ► to move the cursor to the desired location on the graph of the function.

7. **Press ENTER.**

   After pressing ENTER, the slope (derivative) is displayed at the bottom of the screen. This is illustrated in the third graph in Figure 11-5.

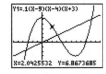
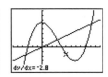

Select function   Enter x Press ENTER

Figure 11-5: Steps in finding the slope at a specified value of x.

Evaluating a Definite Integral

If $f(x)$ is positive for $a \leq x \leq b$, and then the definite integral $\int_a^b f(x)\,dx$ also gives the area between the curve and the x-axis for $a \leq x \leq b$. To evaluate the definite integral, perform the following steps:

1. **Graph the function $f(x)$ in a viewing window that contains the lower limit a and the upper limit b.**

   Graphing a function and setting the viewing window are explained in Chapter 9. To get a viewing window containing a and b, these values must be between **Xmin** and **Xmax.**

2. **Set the Format menu to ExprOn and CoordOn.**

   Setting the Format menu is explained in Chapter 9.

3. **Press 2nd TRACE to access the Calculate menu.**

4. **Press 7 to select the $\int f(x)\,dx$ option.**

5. **If necessary, repeatedly press ▲ until the appropriate function appears at the top of the screen.**

   This process is illustrated in the first graph in Figure 11-6.

6. **Enter the value of the lower limit a.**

   To do so, use the keypad to enter the value of the lower limit a. As you use the keypad, **X=** appears, replacing the coordinates of the cursor location appearing at the bottom of the screen in the previous step. The number you key in appears after **X=**. This is illustrated in the second graph in Figure 11-6. If you make a mistake when entering your number, press CLEAR and re-enter the number.

7. **Press ENTER.**

   After pressing ENTER, a left bound indicator appears at the top of the screen.

8. **Enter the value of the upper limit b and press ENTER.**

   After pressing ENTER, the value of the definite integral appears at the bottom of the screen and the area between the curve and the x-axis, for $a \leq x \leq b$, will be shaded. This is illustrated in the third graph in Figure 11-6.

The shading of the graph produced by using the $\int f(x)\,dx$ option on the Calculate menu does not automatically vanish when you use another Calculate option. To erase the shading, press 2nd PRGM 1 to invoke the **ClrDraw** command on the Draw menu. The graph is then redrawn without the shading.

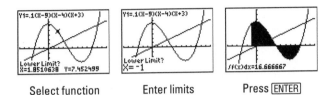

Select function   Enter limits   Press ENTER

Figure 11-6: Steps in evaluating a definite integral.

Chapter 12

Drawing on a Graph

. . . .

In This Chapter

▶ Drawing line segments on a graph

▶ Drawing circles on the graph

▶ Drawing functions on a graph

▶ Writing text on a graph

▶ Freehand drawing on a graph

▶ Saving a picture of your graphs and drawings

. . . .

*A*fter you've graphed your functions, parametric equations, polar equations, or sequences, you can have the calculator draw lines, circles, tangents, and functions on the graph. You can also have the calculator write text on the graph, shade the graph, even save a picture of the graph and the drawings. As an example, Figure 12-1 was constructed using the Draw options provided by the calculator.

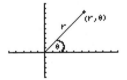

Figure 12-1: An example of drawing on a graph.

Before drawing on a graph, make sure all Y=, Window, Mode, and Format settings are exactly the way you want them to be. If you change any of these settings after you've drawn on the graph, your drawings disappear when the calculator redraws to graph with the new settings.

Drawing Lines, Circles, Tangents, and Functions on a Graph

Have you ever wanted to graph the vertical line $x = a$ but couldn't because the calculator graphs only functions of the form $y = f(x)$? Have you been disappointed by the results you get when you graph the circle $x^2 + y^2 = r^2$ by graphing the functions $y = \sqrt{(r^2 - x^2)}$ and $y = -\sqrt{(r^2 - x^2)}$? And wouldn't it be nice if you could get the calculator to draw a tangent on a graph without your having to tell the calculator the equation of the tangent? If you answered "yes" to any of these questions, then you are in for a treat. The Draw feature on the calculator takes care of all these problems. It will even graph the inverse of a function!

Drawing line segments

You can draw line segments, horizontal lines, and vertical lines on the graph of functions, parametric equations, polar equations, or sequences. To draw a line segment on your graph, follow these steps:

1. **Graph the functions, parametric equations, polar equations, or sequences.**

    For more about graphing these entities, see (respectively) Chapters 9, 13, 15, or 16.

2. **Press 2nd PRGM 2 to select the Line option from the Draw menu.**

    The coordinates of the cursor appear at the bottom of the screen, provided that **CoordOn** is highlighted in the Format menu. (The Format menu is described in Chapter 9.)

3. **Use the ▶◀▲▼ keys to move the cursor to the location of one endpoint of the segment and press ENTER.**

4. **Use the ▶◀▲▼ keys to move the cursor to the location of the other endpoint of the segment and press ENTER.**

5. **Repeat Steps 3 and 4 to draw another segment or press CLEAR when you're finished drawing line segments.**

If you want to erase your drawing or a portion of your drawing, the section in this chapter titled "Erasing Drawings" tells you how to do this.

Drawing horizontal and vertical lines

The procedures for drawing horizontal and vertical lines are similar to the above procedure for drawing a line segment. To draw a horizontal line, perform the following steps. (For a vertical line, you would select the Vertical option from the Draw menu in Step 2, below. Otherwise, the procedures are the same.)

1. **Graph the function, parametric equations, polar equation, or sequence.**

   Graphing such entities is described in Chapters 9, 13, 15, or 16, respectively.

2. **Press** $\boxed{\text{2nd}}\boxed{\text{PRGM}}\boxed{3}$ **to select the Horizontal option from the Draw menu.**

3. **Use the** $\boxed{\blacktriangleright}\boxed{\blacktriangleleft}\boxed{\blacktriangle}\boxed{\blacktriangledown}$ **keys to move the cursor to the location of the horizontal line and press** $\boxed{\text{ENTER}}$.

   If using the $\boxed{\blacktriangleright}\boxed{\blacktriangleleft}\boxed{\blacktriangle}\boxed{\blacktriangledown}$ keys doesn't take you to the exact location of your horizontal line, press $\boxed{\text{2nd}}\boxed{\text{MODE}}$ to exit the graph. Then press $\boxed{\text{2nd}}\boxed{\text{PRGM}}\boxed{3}$, enter the *y*-coordinate of the horizontal line, and press $\boxed{\text{ENTER}}$. Your horizontal line appears on the graph.

4. **Repeat Step 3 to draw another horizontal line or press** $\boxed{\text{CLEAR}}$ **when you're finished drawing horizontal lines.**

   If you used the Tip in Step 3 to draw the horizontal line, then you won't have to press $\boxed{\text{CLEAR}}$ when you're finished. But you must press $\boxed{\text{2nd}}\boxed{\text{PRGM}}\boxed{3}$ again if you want to draw more horizontal lines.

Drawing circles on a graph

The steps for drawing a circle on a graph are similar to the steps in the last section for drawing a line segment on the graph. Press $\boxed{\text{2nd}}\boxed{\text{PRGM}}\boxed{9}$ to select the **Circle** option from the Draw menu, place the cursor at the center of the circle and press $\boxed{\text{ENTER}}$, then move the cursor to a point that defines the radius of the circle and press $\boxed{\text{ENTER}}$. Repeat this process to draw another circle or press $\boxed{\text{CLEAR}}$ when you're finished drawing circles.

Alternatively, if you know the center and radius of the circle, press $\boxed{\text{2nd}}\boxed{\text{MODE}}$ to exit the graph. Then press $\boxed{\text{2nd}}\boxed{\text{PRGM}}\boxed{9}$. Enter the center and radius of the circle, separated by commas, and then press $\boxed{)}\boxed{\text{ENTER}}$, as shown in the second picture in Figure 12-2, where the center of the circle is (–1, 0) and the radius is $\pi/1.8$.

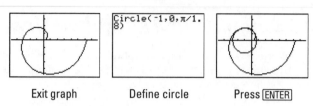

| Exit graph | Define circle | Press ENTER |

Figure 12-2: Steps in drawing a circle on a graph.

Drawing tangents on a graph

You can draw a tangent line on the graph of a function, parametric equation, or polar equation. If you're drawing a tangent to a function, as a bonus, the calculator gives you the equation of that tangent. If you're drawing a tangent to a parametric or polar graph, the calculator will give you the slope of that tangent. To draw a tangent, follow these steps:

1. **Set the Format menu to ExprOn and CoordOn and graph the functions, parametric equations, or polar equations.**

   Graphing such entities and setting the Format menu are described in Chapters 9, 15, or 16, respectively.

2. **Press 2nd PRGM 5 to select the Tangent option from the Draw menu.**

3. **If necessary, repeatedly press ▲ until the cursor is on the appropriate graph.**

4. **Use the ▶ ◀ ▲ ▼ keys to move the cursor to the location of the tangent line and press ENTER.**

   Instead of using the ▶ ◀ ▲ ▼ keys, you can use the keypad to enter the precise location of the tangent. To do so, enter the value of the independent variable, and then press ENTER.

Drawing functions on a graph

To draw a function on a graph, press 2nd PRGM 6 to select the **DrawF** option from the Draw menu. Enter the function as a function of x and press ENTER. If you're in Parametric, Polar, or Sequence mode, the letter x is entered into the calculator by pressing ALPHA STO▸, as shown in Figure 12-3.

If pressing 2nd PRGM 6 places the **DrawF** command after something that was previously enter in the Home screen, press CLEAR to erase that line and then press 2nd PRGM 6.

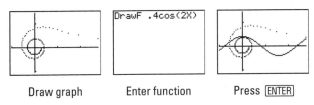

| Draw graph | Enter function | Press ENTER |

Figure 12-3: Steps in drawing a function on a graph.

Drawing the inverse function

You can draw inverses of functions in Function mode only. The mode menu is discussed in Chapter 1. To draw the inverse, press 2nd PRGM 8 to select the **DrawInv** option form the Draw menu. Press VARS ▶ 1 to open the Variables Function menu. Press the number of the function whose inverse you want to graph. Press ENTER and the inverse function is drawn on the graph, as illustrated in Figure 12-4.

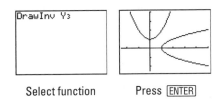

| Select function | Press ENTER |

Figure 12-4: Steps in drawing the inverse of a function.

Shading Between Functions

Regardless what mode the calculator is in, you can shade only between functions written specifically as functions of the independent variable *x*. These functions don't have to be entered in the calculator ahead of time. When you shade the area between two functions, you can shade the whole area or just a portion of that area. To shade functions, follow these steps:

1. **Press** 2nd PRGM 7 **to select the Shade option from the Draw menu.**

2. **Enter the definitions of the lower function, press** . **, and then enter the definition of the upper function.**

   The order in which you enter the functions determines what portion of the graph will be shaded. In the second picture in Figure 12-5, the lower function is Y_1 and the upper function is Y_2. In the third picture in this figure, Y_2 is the lower function.

If your calculator is in Function mode and you have already entered the functions in the calculator, then you can enter them into the Shade command by using their function names. To do so, press $\boxed{\text{VARS}}\boxed{\blacktriangleright}\boxed{1}$ to open the Variables Function menu. Then press the number of the function you want to enter in the Shade command, as shown in the first picture in Figure 12-5.

If you haven't entered the functions in the calculator (or the calculator isn't in Function mode), enter the function definitions in the Shade command. For example, the first picture in Figure 12-5 is equivalent to Shade($x^2 - 1, x + 1$). In parametric, polar, or sequence mode, you enter the letter x into the calculator by pressing $\boxed{\text{ALPHA}}\boxed{\text{STO}\blacktriangleright}$.

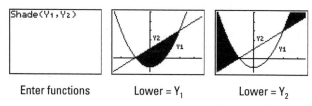

Enter functions   Lower = Y_1 Lower = Y_2

Figure 12-5: Shading the area between two functions.

If you're shading the whole area between two functions and you like the default shading (shown in Figure 12-5), then after completing Step 2, press $\boxed{)}\boxed{\text{ENTER}}$ and skip the remaining steps.

3. **Press $\boxed{,}$.**

4. **Enter the minimum value of x in the shaded area, press $\boxed{,}$, and then enter the minimum value of x in the shaded area.**

The minimum and maximum values of x must be between **Xmin** and **Xmax** in the Window editor. If they are not, you won't see the total shaded area on the calculator.

If you're shading a portion of the area between two functions you defined in Step 4, and you like the default shading (shown in Figure 12-5), then press $\boxed{)}\boxed{\text{ENTER}}$ and skip the remaining steps.

5. **Press $\boxed{,}$.**

6. **Enter a number, 1 through 4, of the type of shading you want, and then press $\boxed{,}$.**

Enter 1 for vertical line shading, 2 for horizontal line shading, 3 for negatively sloping diagonal line shading, or 4 for positively diagonal line shading.

7. **Enter a number from 1 through 8 to set the resolution of the shading, and then press ⎵.**

   If you enter 1, every pixel on the screen is shaded, as shown in Figure 12-5. If you enter 2, every *other* pixel is shaded, and so on. My favorite number for the resolution is 4, which shades every fourth pixel (as in Figure 12-6).

8. **Press ENTER to shade the graph.**

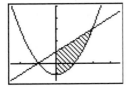

Figure 12-6: Shade (Y_1, Y_2, 0, 2, 3, 4).

Writing Text on a Graph

Writing text on a graph is quite useful. For example, you can indicate the numerical value of tick marks on the axes and you can label the graphs. On the other hand, writing text on a graph can be very frustrating because the calculator has no undo feature and it takes practice to figure out where to place the text so that it doesn't interfere with the graph. To write text on the graph, follow these steps:

1. **Graph the functions, parametric equations, polar equations, or sequences.**

   Graphing such entities is described in Chapters 9, 13, 15, or 16, respectively.

2. **Press 2nd PRGM 0 to select the Text option from the Draw menu.**

3. **Position the cursor on the screen at the spot where you want to start writing text.**

   Finding this spot takes practice. The cursor looks like a very small + sign and the text you write will be about three times as tall as the cursor and two and a half times as wide. The upper-left corner of the text will be positioned in the fourth quadrant of this + sign.

4. **Enter your text.**

   Use the ALPHA key to enter letters and the number keys to enter numbers. The = sign is entered by pressing 2nd MATH 1 and the arithmetic functions are entered by pressing ⊞, ⊟, ⊠, or ⊡. Other useful symbols can be found in the Catalog. For example, the small letter *r* in Figure 12-1 was entered by pressing 2nd 0 to enter the Catalog, repeatedly pressing ▲ until the indicator was next to *r*, and then pressing ENTER.

If you don't like the location of the text you've placed on the screen, move the cursor to the upper-left corner of the text and repeatedly press ALPHA 0 to replace the text with a space. Doing so erases the text; it may also erase a small part of the graph. The next section tells you how to patch any holes created by using this technique to erase text.

5. **Move the cursor to a new location to enter more text or press CLEAR when you're finished writing text.**

Freehand Drawing on a Graph

You can use the Pen option to place freehand drawings on a graph. The angle symbol around θ in Figure 12-1 was constructed using the Pen. And if using the previous tip left a hole in the graph or on an axis, you can patch it by using the Pen. To use the Pen to create a freehand drawing, perform the following steps:

1. **Graph the functions, parametric equations, polar equations, or sequences.**

   Graphing such entities is described in Chapters 9, 13, 15, or 16, respectively.

2. **Press 2nd PRGM ▲ ENTER to select the Pen option from the Draw menu.**

3. **Move the cursor to the spot where you want to place your free-hand drawing and press ENTER.**

4. **Use the ▶ ◀ ▲ ▼ keys to construct your drawing and press ENTER when you're finished.**

   As you press the ▶ ◀ ▲ ▼ keys, the pixels on the screen are shaded along the path of the cursor.

5. **Repeat Steps 3 and 4 to create another drawing or press CLEAR when you're finished using the pen.**

Erasing Drawings

Because the Draw menu has no Undo feature, the capability to erase is important — and three types of erasing are possible: erase all drawings, erase a *portion* of a drawing or graph, or erase one or more *points* from a drawing or graph. (Erasing points from a graph comes in handy when you really want to emphasize that a function is undefined at a certain point.)

✔ **To erase all drawings:** Press [2nd][PRGM][1] to select **ClrDraw** from the Draw menu. If the graph is displayed at the time you pressed these keys, the graph will be redrawn without the drawings. If you press these keys from the Home Screen, you also have to press [ENTER] to put the **ClrDraw** command into action.

✔ **To erase a portion of a drawing or graph:** The space key ([ALPHA][0]), which is entered as text, can be used to erase a portion of a drawing or graph. Each time you use this technique, it erases an area about three times as high as the cursor and two times as wide. (The previous section, "Writing Text on a Graph," tells you how to enter a space as text.)

✔ **To erase one or more points from a drawing or graph:** Press [2nd][PRGM][▶][2] to select **Pt-Off** from the Draw Points menu. Move the cursor to the point you want to erase and press [ENTER]. Move the cursor to the next point you want to erase and press [ENTER]. When you're finished erasing points, press [CLEAR].

Saving Graphs and Drawings

A graph that contains drawings can be saved only as a Picture. When you recall that Picture from memory, all you can do is look at it. You cannot, for example, change the viewing window or trace the graph.

To Save a graphs and drawings as a Picture, follow these steps:

1. **Press [2nd][PRGM][▶][▶] to access the Draw Store menu.**

2. **Press [1] to store your graph as a Picture.**

3. **Enter an integer from 0 through 9.**

   The calculator can store up to 10 Pictures. If, for example, you enter the number 5, your Picture is stored in the calculator as **Pic5**.

   If you save your Picture as **Pic5** without realizing that you had previously stored another Picture in **Pic5**, the calculator will, without warning or asking your permission, erase the old **Pic5** and replace it with the new **Pic5**. To see a list of the Pictures already stored in your calculator, press [2nd][+][2][8].

If you already have ten Pictures stored in your calculator and don't want to sacrifice any of them, consider saving some of them on your PC (Chapter 22 describes how to do so).

4. **Press** ENTER.

If you want to delete a Picture from the calculator, just perform the following steps:

1. **Press** 2nd + **to access the Memory menu.**

2. **Press** 2 **to access the Mem Mgt/Del menu.**

3. **Press** 8 **to access the Pic files stored in the calculator.**

4. **If necessary, repeatedly press** ▾ **to move the indicator to the Pic you want to delete.**

5. **Press** DEL.

6. **Press** 2nd MODE **to exit this menu and return to the home screen.**

And to recall a Picture, all you have to do is:

1. **Press** 2nd PRGM ▸ ▸ **to access the Draw Store menu.**

2. **Press** 2 **to recall your Picture.**

3. **Enter the number of your stored Picture.**

4. **Press** ENTER.

Part V

Sequences, Parametric Equations, and Polar Equations

The 5th Wave   By Rich Tennant

ROOM 101

"I failed her in algebra but was impressed with the way she animated her equations to dance across the screen, scream like hyenas, and then dissolve into a clip art image of the La Brea Tar Pits."

In this part...

This part shows how to graph sequences, parametric equations, and polar equations. You get a look at how to trace a graph, create tables, and save your graph for future use. And I show you how to convert between rectangular and polar coordinates.

Chapter 13

Graphing Sequences

· ·

In This Chapter

▶ Entering sequences into the calculator

▶ Graphing sequences

▶ Graphing two or three sequences

▶ Saving sequence graphs for future use

▶ Drawing on a sequence graph

· ·

A sequence is simply an ordered list of terms or numbers. The most famous sequence is perhaps the Fibonacci sequence, $\{0, 1, 1, 2, 3, 5, 8, 13, \ldots\}$, where the first two terms are given and the remaining terms are found by adding the previous two terms. The formula used to generate a sequence is referred to by various names, such as a recurrence relation, a recursive function, or an iterative function. Texas Instruments calls these formulas *sequence functions*, so I use this term as well.

In spite of the power of the calculator, it's rather limited when dealing with sequence functions. It can accommodate only three sequence functions, and each function can address only the previous two terms of that function (or of another function). So the calculator can handle the Fibonacci sequence, $u(n) = u(n-1) + u(n-2)$, but nothing fancier.

Entering a Sequence

The following lists the steps for entering a sequence into the calculator. For the sake of simplicity, I tell you how to enter the sequence $u(n)$. But what is stated in these steps for $u(n)$ also applies to the sequences $v(n)$ and $w(n)$.

1. **Press** MODE **and put the calculator in Sequence mode, as shown in Figure 13-1.**

   To highlight an item in the Mode menu, use the ▶◀▲▼ keys to place the cursor on the item, and then press ENTER. Highlight **Seq** in the fourth line to put the calculator in Sequence mode. Highlighting **Float** in the second line displays numbers with as many decimal places as the calculator can handle. If your sequence deals with money, highlight the number **2** in the second line to make the calculator round all numbers to two decimal places. (For more information on the other items on the Mode menu, please refer to Chapter 1.)

Figure 13-1: Setting Sequence mode.

2. **Press** Y= **, enter a value for *n*Min, and press** ENTER **.**

   *n***Min** denotes the first value of the independent variable n in the sequence u(n) and in all other sequences you enter into the calculator. You usually want to set it equal to 1 so your sequences look like {u(1), u(2), u(3), . . .}, {v(1), v(2), v(3), . . .}, and {w(1), w(2), w(3), . . .}. But if your sequences represent an experiment that starts at "time zero," you would want to set *n***Min** equal to 0. This way your sequences would look like {u(0), u(1), u(2), . . .} and so on.

   To enter the number you've chosen for *n***Min**, place the cursor on the number appearing after *n***Min**, press the number keys to enter your new value, and then press ENTER.

   *n***Min** must be set equal to 0 or a positive integer. The calculator isn't equipped to handle negative values for *n***Min**. Setting *n***Min** equal to something other than 0 or a positive integer results in an error message.

3. **Enter the definition of the sequence u(*n*) and press** ENTER **.**

To erase an entry that appears after u(*n*), use the 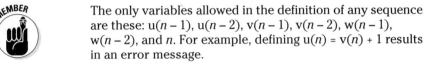 keys to place the cursor to the right of the equal sign and press CLEAR. Then enter your definition for the new sequence.

The sequence function names **u**, **v**, and **w** appear on your keypad in yellow, above ⑦, ⑧, and ⑨, respectively. To enter **u**, for example, press 2nd⑦. To enter the independent variable *n*, press X,T,θ,*n*.

The only variables allowed in the definition of any sequence are these: u(*n* – 1), u(*n* – 2), v(*n* – 1), v(*n* – 2), w(*n* – 1), w(*n* – 2), and *n*. For example, defining u(*n*) = v(*n*) + 1 results in an error message.

4. Enter the value of u(*n*Min) and press ENTER.

u(*n*Min) is left blank, set equal to the first term in the sequence u(*n*), or set equal to the first two terms in u(*n*). It all depends of how the sequences **u**, **v**, and **w** are defined.

The following tells you what value to assign to **u(*n*Min)** and how it's entered in the calculator:

- **u(*n*Min)** can be left blank if none of the sequences **u**, **v**, or **w** use u(*n* – 1) or u(*n* – 2) in their definitions. If **u(*n*Min)** has previously been assigned a value, you can get rid of that value by using the 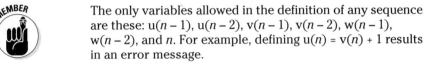 keys to place the cursor after the equal sign, pressing CLEAR to erase it, and then blanking it by pressing ENTER.

- **u(*n*Min)** is set equal to the first term in the sequence u(*n*) if any of the sequences **u**, **v**, or **w** use u(*n* – 1) in their definitions, but none of them use u(*n* – 2). To set **u(*n*Min)** equal to the first term in the sequence u(*n*), use the 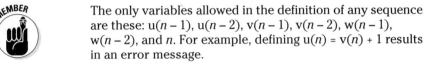 keys to place the cursor after the equal sign, then use the number keys to enter the value you want to assign **u(*n*Min)**, and press ENTER. As you enter this number, the calculator automatically erases any previous value assigned to **u(*n*Min)**; after you press ENTER, the calculator automatically places curly braces around the number you just entered.

- **u(*n*Min)** is set equal to the first two terms in the sequence u(*n*) if any of the sequences **u**, **v**, or **w** use u(*n* – 2) in their definitions. And to complicate matters, these terms must be entered in reverse order, and you must supply the curly braces. You enter these terms after the equal sign for **u(nMin)** by keying in: {second term in u(n), first term in u(n)}. The curly braces are entered into the calculator by pressing [2nd][(] and [2nd][)]. (An example appears at the bottom of Figure 13-2.)

If you're ever in doubt about how to set **u(nMin)**, you can never go wrong by setting it equal to the first two terms in the sequence u(n). However, it's not always mathematically easy to *find* these two terms.

In the example in Figure 13-2, **u(nMin)** is left blank because none of the sequences **u**, **v**, or **w** used u(n – 1) or u(n – 2) in their definitions. **v(nMin)** is set equal to the first term in v(n) because v(n) used v(n – 1) in its definition. In this example **v(nMin)** was assigned the value of 5. Because u(n) used w(n – 2) in its definition, **w(nMin)** is set equal to the first two terms in w(n), listed in reverse order.

Figure 13-2: An example of entering sequences.

A comment on setting u(nMin)

Figure 13-2 illustrates a situation of when you may want to, but don't have to, deviate from the requirements for setting **u(nMin)**. In this example, the calculator does not know the value of the first two terms of u(n), nor can it evaluate them. After all, u(1) = w(-1) and u(2) = w(0), but the calculator can evaluate w(n) only for $n \geq 1$.

Luckily, the calculator does not have to know these terms. It just calculates u(n) for $n \geq 3$. When you graph u(n), or produce a table of the values for u(n), the first two terms of u(n) don't appear, but the rest of the sequence does. If you can't live with the omission of these terms in the graph or in the table, then set **u(nMin)** equal to the first two terms in u(n), listed in reverse order.

In this example, u(n) = w(n – 2) + n = [2(n – 2) + 1] + n = 3n – 3. So evaluating the first two terms of u(n) isn't that difficult. However, for more complicated sequences, it may not be worth your time to figure out the values of these terms. After all, didn't you get the calculator so you wouldn't have to do scads of math by hand? In situations like this, I recommend just living with the omission of the first two terms in the graph and the table.

Graphing Sequences

After you have entered the sequence functions into the calculator, as described in the previous section, you can use the following steps to graph the sequences:

1. **Turn off any Stat Plots that you don't want to appear in the graph of your sequences.**

    The first line in the Y= editor tells you the graphing status of the Stat Plots. (Stat Plots are discussed in Chapter 19.)

    If **Plot1**, **Plot2**, or **Plot3** is highlighted, then that Stat Plot will be graphed along with the graph of your sequences. If it's not highlighted, it won't be graphed. In Figure 13-2, **Plot1** will be graphed along with the sequences

    To turn off a highlighted Stat Plot in the Y= editor, use the ▶◀▲▼ keys to place the cursor on the highlighted Stat Plot and then press ENTER. If you want to graph the Stat Plot later on, use this same process to rehighlight it.

    When you're graphing sequences, Stat Plots that are turned on when you don't really want them graphed cause problems. The most common problem is the ERR: INVALID DIM error message, which gives you precious little insight into what's causing the problem. So if you aren't planning to graph a Stat Plot along with your sequences, make sure all Stat Plots are turned off. (The previous paragraph tells you how to turn them off.)

2. **Press 2nd ZOOM to access the Format menu.**

3. **Set the format for the graph in the first line of the Format menu by placing the cursor on the desired format and then pressing ENTER.**

    Figure 13-3 pictures the Format menu when the calculator is in Sequence mode. The second line of this menu gives you a choice: You can display the coordinates of a point in rectangular form (**RectGC**) or polar form (**PolarGC**). The first line offers various formats for graphing sequences. Assuming you've selected **RectGC**, here is what the calculator graphs in each format on the first line of the menu:

    - **Time:** This is the most common format for graphing sequences because it graphs the sequences as a function of the independent variable n. That is, in the Time format, the points $(n, u(n))$, $(n, v(n))$, and $(n, w(n))$ are graphed.

- **Web:** This format produces a *web plot* (also known as a *cobweb plot* because of its shape) for a sequence u(*n*). Use it when you want to see whether u(*n*) converges to an equilibrium point or just veers off into space. It graphs the points (u(*n*), u(*n* + 1)) and the line *y* = *x*. Using the TRACE and ▶ keys then shows you whether the sequence is converging or diverging. This tracing process is explained in Chapter 14.

In Web format, the calculator places two restrictions on how the sequence u(*n*) is defined. First, it requires that u(*n* – 1) appear as a variable in the definition of u(*n*); second, it requires that u(*n* – 1) be the only variable used in this definition. Defining u(*n*) as u(*n*) = u(*n* – 1) + u(*n* – 2) or as u(*n*) = v(*n* – 1) results in an error message. And because u(*n*) = u(*n* – 1) + *n* uses the variable *n* in its definition, it too results in an error message when it's used in the Web format.

- **uv, vw, and uw:** Use these formats to create phase plots when you want to see how one sequence affects another sequence. The **uv** format graphs the points (u(*n*), v(*n*)); the **vw** format graphs the points (v(*n*), w(*n*)); **uw** format graphs the points (u(*n*), w(*n*)).

The remaining items on the Format menu are explained in Chapter 9. When you're graphing sequences, there's no harm in leaving these items highlighted as they appear in Figure 13-3.

Figure 13-3: Format menu in Sequence mode.

4. **Press WINDOW to access the Window editor.**

5. **After each of the first four window variables, enter a numerical value that is appropriate for the sequences you're graphing. Press ENTER after entering each number.**

Figure 13-4 pictures the Window editor when the calculator is in Sequence mode. The following list explains the variables you must set in this editor:

- **nMin:** This setting contains the first value of the independent variable n. It must be set equal to 0 or a positive integer.

  nMin appears both in the Window editor and in the Y= editor. If you change the value of **nMin** in either the Window editor or the Y= editor, the value of **nMin** will automatically be updated in both editors. (A more detailed explanation of **nMin** appears in Step 2 of the previous section.)

```
WINDOW
 nMin=1
 nMax=100
 PlotStart=1
 PlotStep=1
 Xmin=0
 Xmax=100
↓Xscl=5
```

Figure 13-4: Window editor in Sequence mode.

- **nMax:** This setting contains the largest value of the independent variable n that you want the calculator to use to evaluate your sequences. For example, if **nMin** = 0 and **nMax** = 100, the calculator will evaluate the first 101 terms in each sequence.

  nMax must be set equal to a positive integer that is greater than **nMin**. Be reasonable when entering a value for **nMax**. If you're graphing three sequences and **nMax** = 1,000, it's going to take the calculator at least a minute to get the job done.

If it's taking a long time for the calculator to graph your sequences, and you start to regret that rather large value you placed in **nMax**, press [ON] to terminate the graphing process. You can then go back to the Window editor and adjust the **nMax** setting.

- **PlotStart:** This setting tells the calculator where you want to start graphing in each sequence. For example, if **PlotStart** = 20, then the calculator starts the graph with the 20th term in each sequence.

This isn't always intuitive. If **nMin** = 1, then the 20th term in the sequence u(n) is u(20). But if **nMin** = 0, then the 20th term in this sequence is u(19).

PlotStart must be set equal to a positive integer. Usually it's set equal to 1 so that the graph starts at the beginning of a sequence.

• **PlotStep:** This setting tells the calculator if you want to skip graphing any terms in each sequence. A setting of 1 tells the calculator to graph every term in each sequence; a setting of 2 tells it to graph every other term; and a setting of 3 tells it to graph every third term, and so on.

PlotStep must be set equal to a positive integer. Usually it's set equal to 1 so that the graph shows all terms in your sequences.

You use the remaining items in the Window editor to set a viewing window for the graph (a procedure explained in detail in Chapter 9). If you know the dimensions of the viewing window required for your graph, go ahead and assign numerical values to the remaining variables in the Window editor and advance to Step 8. If you don't know the minimum and maximum *y* values required for your graph, the next step tells you how to get the calculator to find them for you.

6. **Press** ZOOM 0 **to access ZoomFit.**

After you have assigned values to *n***Min**, *n***Max**, **PlotStart**, and **PlotStep**, **ZoomFit** determines appropriate values for **Xmin**, **Xmax**, **Ymin**, and **Ymax** and graphs your sequences. **ZoomFit** graphs the sequences in the smallest possible viewing window, however, and it won't assign appropriate values to **Xscl** and **Yscl**.

If you like the way the graph looks, then you can skip the remaining steps. If you'd like spruce up the graph, Step 7 gives you some pointers.

7. **Press** WINDOW **and adjust the values assigned to the X and Y settings. Press** ENTER **after entering each new number.**

Here's how to readjust the viewing window after using **ZoomFit**:

• **Xmin:** If you want to see the *y*-axis on your graph, set **Xmin** equal to 0.

• **Xmax:** If you're interested in seeing a more detailed graph of the beginning of the sequence, decrease the value of **Xmax**.

• **Xscl:** Set this equal to a value that doesn't make the *x*-axis look like railroad tracks — that is, an axis with far too many tick marks. Twenty or fewer tick marks makes for a nice looking axis.

- **Ymin and Ymax:** If you don't want the graph to hit the top and bottom of the screen, decrease the value assigned to **Ymin** and increase the value assigned to **Ymax**. If you want to see the *x*-axis on the graph, assign values to **Ymin** and **Ymax** so that zero is strictly between these two values.

- **Yscl:** Set this equal to a value that doesn't make the *x*-axis look like railroad tracks (too many tick marks). Fifteen or fewer tick marks is a nice number for the *y*-axis.

8. Press GRAPH to regraph the sequences.

When you're graphing sequences, if you get the ERR: INVALID error message and then select the **Goto** option, the calculator sends you to the definition of the "problem" sequence that you placed in the calculator. More often than not, however, the true location of the problem is in the way you defined **u(*n*Min)**, **v(*n*Min)**, or **w(*n*Min)**, and not in the way you defined the *sequence* (as the calculator leads you to believe). For example, if $w(n) = u(n - 2) + v(n - 1)$ and the calculator places the cursor after $u(n - 2)$ when you select the **Goto** option, then the problem is most likely caused by not setting **u(*n*Min)** equal to the first two terms in u(*n*). The previous section explains how to set **u(*n*Min)**.

Graphing Several Sequences

Even if you've put the calculator in the Sequential mode for graphing functions (as indicated in the sixth line of Figure 13-1), it's going to graph all sequences simultaneously. So you can't tell the sequences apart simply by looking at the order in which they're graphed.

To identify the graphs of several sequences, set a different graph style for each sequence:

1. Press Y= to access the Y= editor.

2. Use the ▶◀▲▼ keys to place the cursor on the icon appearing at the very left of the definition of the sequence.

3. Repeatedly press ENTER until you get the desired graph style.

When the calculator graphs the discrete points in a scatter plot, it gives you a choice of graphing them as dots, little plus signs, or small squares. It's a shame that you don't get the same choices when graphing the discrete points in a sequence. Unfortunately, the graphing styles for sequences are Dotted Line, Line, and Thick Line. (Go figure!)

Each time you press $\boxed{\text{ENTER}}$, you get a different style. In Figure 13-2, u(n) is set to the default Dotted Line style, whereas v(n) and w(n) are respectively set to the Line and Thick Line styles.

TIP If you don't want the calculator to graph one of your sequences, un-highlight its equal sign. To graph it at a later time, rehighlight its equal sign. You can do so in the Y= editor by using the $\boxed{\blacktriangleright}\boxed{\blacktriangleleft}\boxed{\blacktriangle}\boxed{\blacktriangledown}$ keys to place the cursor on the equal sign in the definition of the sequence and then pressing $\boxed{\text{ENTER}}$ to toggle the equal sign between highlighted and unhighlighted. In the example in Figure 13-2, the calculator won't graph sequence w(n), but it will graph u(n) and v(n).

Saving a Sequence Graph

After you've graphed your sequences (as described in the last two sections), you can save the graph and its entire Window, Y= editor, Mode, and Format settings in a Graph Database. When you recall the Database at a later time, you get more than just a picture of the graph. The calculator also restores the Window, Y= editor, Mode, and Format settings to those stored in the Database. Thus you can, for example, trace the recalled graph.

The procedures for saving and recalling a graph database are described in Chapter 9.

Drawing on a Sequence Graph

After you've graphed your sequences (as described earlier in this chapter), you can draw lines and functions on the graph, write text on it, and save a picture of the graph and the drawings. This capability is handy if you want to illustrate the way a sequence converges to a number, or show that a sequence can be approximated by a function.

The procedures for drawing on a graph and for saving the result as a picture are described in Chapter 12.

Chapter 14

Exploring Sequences

In This Chapter

▶ Using Zoom commands with sequences

▶ Tracing the graph a sequence

▶ Constructing a table of sequence values

▶ Evaluating sequences

Exploring Sequence Graphs

When you explore the graph of a sequence, you're usually interested in seeing where the sequence goes. Does it level off and converge to a number? Does it veer off into space? What does it look like from the 50th term on? What's the value of the 100th term? The calculator's Zoom and Trace features help you answer such questions.

Using ZOOM in Sequence mode

Pressing ZOOM accesses the Zoom menu. On this menu you see all the Zoom commands that are available for graphing functions. Not all of them are useful when graphing sequences, however. Here are the ones that I use when graphing sequences:

 ✔ **ZoomFit:** This command finds a viewing window for a specified portion of the sequence. How to use **ZoomFit** is explained in Chapter 13.

 ✔ **Zbox:** After the graph is drawn (as described in Chapter 13), this command allows you to define a new viewing window for a portion of your graph by enclosing it in a box as illustrated in Figure 14-1. The box becomes the new viewing window. To construct the box:

   1. Press ZOOM 1.

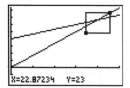

Figure 14-1: Using **ZBox.**

2. **Define a corner of the box.**

   To do so, use the ▶◀▲▼ keys to move the cursor to the spot where you want one corner of the box to be located, and then press ENTER. The calculator marks that corner of the box with a small square.

3. **Construct the box.**

   To do so, press the ▶◀▲▼ keys. As you press these keys, the calculator draws the sides of the box.

   When you use **ZBox**, if you don't like the size of the box, you can use any of the ▶◀▲▼ keys to resize the box. If you don't like the location of the corner you anchored in Step 2, press CLEAR and start over with Step 1.

4. **When you're finished drawing the box, press** ENTER **and the graph will be redrawn in the viewing window specified by the box.**

   When you use **Zbox,** ENTER is pressed only two times. The first time you press it is to anchor a corner of the zoom box. The next time you press ENTER is when you're finished drawing the box and are ready to have the calculator redraw the graph.

✔ **Zoom In and Zoom Out:** After the graph is drawn (as described in Chapter 13), these commands allow you to zoom in on a portion of the graph or to zoom out from the graph. They work very much like a zoom lens on a camera. To use these commands:

1. **Press** ZOOM 2 **if you want to zoom in, or press** ZOOM 3 **if you want to zoom out.**

2. **Use the** ▶◀▲▼ **keys to move the cursor to the spot on-screen from which you want to zoom in or zoom out.**

This spot becomes the center of the new viewing window.

3. Press ENTER.

You can press ENTER again to zoom in closer or to zoom out one more time.

4. Press CLEAR **when you're finished zooming in or zooming out.**

If you used a Zoom command to redraw a graph and then want to undo what that command did to the graph, press ZOOM ▶ 1 to select **ZPrevious** from the Zoom Memory menu. The graph is then redrawn as it appeared in the previous viewing window.

Tracing a sequence

After you have graphed your sequences (as described in Chapter 13), press TRACE and then use ▶ to investigate the sequence or sequences. Here's what you will see, and what you can do to change things:

✔ **The definition of the sequence:** The definition of the sequence you're tracing is displayed in the upper-left corner of the screen, provided the calculator is in **ExprOn** format (refer to Figure 13-3 in Chapter 13). If you have graphed more than one sequence and you would like to trace a different sequence, press ▲. Each time you press this key, the cursor jumps to another sequence. Eventually it jumps back to the original sequence.

✔ **The independent variable *n*:** The value of *n* corresponding to the cursor location is displayed in the lower-left corner of the screen, provided the calculator is in **CoordOn** format (refer to Figure 13-3 in Chapter 13). When you press TRACE, the cursor is placed at the beginning of the graph of the sequence and *n* displays the value you assigned to *n***Min** in the Window editor. Each time you press ▶, the cursor moves to the next plotted point in the graph — and the value of *n* changes to the value of the independent variable corresponding to that plotted point.

If you press ◀, the cursor will move left to the previously plotted point in the sequence. And if you press ▲ to trace a different sequence, the tracing of that sequence will start at the same value of *n* that was displayed on-screen before you pressed this key.

When you're using TRACE, if you want to start tracing your sequence at a specific value of the independent variable *n*, just key in that value and press ENTER. (The value you assign to *n* cannot be greater than *n*Max; if it is, you will get an error message.) After you press ENTER, the trace cursor moves to the point on the graph corresponding to that value of *n*. But the calculator does not change the viewing window. So if the value you assigned to *n* is greater than **Xmax**, you won't see the cursor; it's on the part of the graph outside the viewing window. The sidebar "Panning in Sequence mode" tells you how to get the cursor and the graph in the same viewing window.

✔ **The values of *x* and *y*:** At the bottom of the screen you see the values of the *x*- and *y*- coordinates of the cursor location, provided the calculator is in **CoordOn** format (see Figure 13-3 in Chapter 13). In the **PolarGC** format, the coordinates of this point display in polar form. The relationship between these coordinates and the sequence depends on how you set the format for the sequence. Setting the format is explained in Chapter 13. Here's what you will see in the various formats:

• **Time format:** In this format the *x*-coordinate is the independent variable *n* and the *y*-coordinate is the corresponding value of the sequence at *n*.

• **In the Web format:** In this format, the trace cursor starts on the *x*-axis at the first term in the sequence u(*n*). When you press ▶, the cursor moves vertically to the graph of the points (u(*n*), u(*n* + 1)). And when you press ▶ again, it moves horizontally to the line *y* = *x*. Each time you press ▶ after that, the cursor repeats this alternating vertical and horizontal movement. In addition, a vertical or horizontal line connects all points traced by the cursor. This is illustrated in Figure 14-2.

In Web format, if *n*Min =1, the points traced by the trace cursor are (u(1), 0), (u(1), u(2)), (u(2), u(2)), (u(2), u(3)), (u(3), u(3)), (u(3), u(4)), and so on. You use this format when you want to see if a sequence converges or diverges. Figure 14-2 shows an example of using the Web format and TRACE to investigate the convergence of a sequence.

• **uv, vw, and uw formats:** In the **uv** format, the
x-coordinate is the value of u(*n*) and the y-coordinate is
the value of v(*n*). In the **vw** format, the x-coordinate is the
value of v(*n*) and the y-coordinate is the value of w(*n*).
And in the **uw** format, the x-coordinate is the value of
u(*n*) and the y-coordinate is the value of w(*n*).

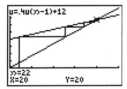

Figure 14-2: Tracing a convergent sequence in Web format.

Press $\boxed{\text{CLEAR}}$ to terminate tracing the graph. This also removes the
name of the function and the coordinates of the cursor from the
screen.

Panning in Sequence mode

When you're tracing a sequence and the cursor hits an edge of the screen, if *n***Max**
is greater than **Xmax** and you continue to press $\boxed{\blacktriangleright}$, the coordinates of the cursor are
displayed at the bottom of the screen. However, the calculator does not automati-
cally adjust the viewing window; you won't see the cursor on the graph. To rectify
this situation, make a mental note of the value of *n* (or jot it down) and then press
$\boxed{\text{ENTER}}$. The calculator then redraws the graph centered at the location of the cursor
at the time you pressed $\boxed{\text{ENTER}}$.

Unfortunately, after the graph is redrawn, the trace cursor is placed at the beginning of the first sequence appearing in the Y= editor. To get the trace cursor back
on the part of the graph displayed in the new viewing window, key in that value of
n that you made note of, and then press $\boxed{\text{ENTER}}$. If you weren't tracing the first
sequence appearing in the Y= editor, use $\boxed{\blacktriangle}$ to place the cursor on the graph you
were tracing. The trace cursor then appears in the middle of the screen and you
can use $\boxed{\blacktriangleright}$ or $\boxed{\blacktriangleleft}$ to continue tracing the sequence.

> If the name of the sequence and the value of the independent vari-
> able *n* are interfering with your view of the graph when you use
> TRACE, increase the height of the window by pressing WINDOW and
> decreasing the value of **Ymin** and increasing the value of **Ymax**.

Displaying Sequences in a Table

After you've entered the sequence functions into the calculator (as
described in Chapter 13), you can have the calculator create a table
of the terms in the sequences. To create a table, follow these steps:

**1. Highlight the equals sign of those sequences in the Y=
editor that you want to appear in the table.**

Only those sequences in the Y= editor that are defined with
a highlighted equal sign will appear in the table. To highlight
or un-highlight an equal sign, press Y=, use the ▶◀▲▼
keys to place the cursor on the equal sign in the definition
of the sequence, and then press ENTER to toggle the equal
sign between highlighted and un-highlighted.

2. Press 2nd WINDOW to access the Table Setup editor.

3. Enter a number in TblStart and then press ENTER.

TblStart is the first value of the independent variable *n* to
appear in the table. In Figure 14-3, **TblStart** is assigned the
value 1.

When constructing a table of sequence values, be sure to
set **TblStart** equal to 0 or to a positive integer.

To enter the number you've chosen for **TblStart**, place the
cursor on the number appearing after the equal sign, press
the number keys to enter your number, and then press
ENTER.

Figure 14-3: The Table Setup editor.

4. Enter a number in ΔTbl and then press ENTER.

ΔTbl gives the increment for the independent variable *n*. For example, if **TblStart** = 1 and **ΔTbl** = 5, then the table for sequence u(*n*) will contain the values of u(1), u(6), u(11), and so on. In Figure 14-3, **ΔTbl** is assigned the value 1.

When constructing a table of sequence values, **ΔTbl** must be set equal to a positive integer.

To enter the number you have chosen for **ΔTbl**, place the cursor on the number appearing after the equal sign, press the number keys to enter your number, and then press [ENTER].

5. Set the mode for Indpnt and Depend.

To change the mode of either **Indpnt** or **Depend**, use the [▶][◀][▲][▼] keys to place the cursor on the desired mode, either **Auto** or **Ask**, and then press [ENTER].

To have the calculator automatically generate the table for you, put both **Indpnt** and **Depend** in **Auto** mode. The first table in Figure 14-4 was constructed this way.

If you want to create a user-defined table in which you specify which values of the independent variable *n* are to appear in the table and then have the calculator figure out the corresponding sequence values for you, put **Indpnt** in **Ask** mode and **Depend** in **Auto** mode. How you construct the table is explained in Step 6. The second table in Figure 14-4 was constructed this way.

For a user-defined table, you don't have to assign values to **TblStart** and **ΔTbl** in the Table Setup editor.

The other combinations of mode settings for **Indpnt** and **Depend** are rather useless when you're dealing with sequences.

n	u(*n*)	v(*n*)
1	0	5
2	3	6
3	6	7
4	9	8
5	12	9
6	15	10
7	18	11

n=1

n	u(*n*)	v(*n*)
5	12	9
20	57	24
50	147	54

n=

Automatically   User-defined
generated

Figure 14-4: Tables of sequences.

6. Press [2nd][GRAPH] to display the table.

When you display the table, what you see on-screen depends on the modes you set for **Indpnt** and **Depend** in Step 5. And what you can do with the table also depends on these modes. Here's what you see and what you can do:

- **An automatically generated table:**

    If **Indpnt** and **Depend** are both in **Auto** mode, then when you press [2nd][GRAPH], the table is automatically generated. To display rows in the table beyond the last row on-screen, repeatedly press [▼] until they appear. And repeatedly press [▲] to display rows above the first row on-screen.

    Each time the calculator redisplays a table with a different set of rows, it also automatically resets **TblStart** to the value of n appearing in the first row of the newly displayed table. To return the table to its original state, press [2nd][WINDOW] to access the Table Setup editor, and then change the value the calculator assigned to **TblStart**.

    If you're constructing a table for all three sequences, only the first two sequences — $u(n)$ and $v(n)$ — appear on-screen. To see the values in the column for sequence $w(n)$, repeatedly press [▶] until that column appears. Doing so, however, makes the $u(n)$ column disappear. To see it again, repeatedly press [◀] until it appears.

- **A user-generated table:**

    If you put **Indpnt** in **Ask** mode and **Depend** in **Auto** mode so that you can generate your own table, then when you display the table, it should be empty. If it's not, clear the table (as described in the next section, "Clearing a User-Defined Table").

    In an empty table, key in the first value of the independent variable n you want to appear in the table, and then press [ENTER]. The corresponding sequence values automatically appear. Key in the next value of n you want in the table and press [ENTER], and so on. The values for n that you place in the first column don't have to be in any specific order, but they do have to be non-negative integers.

If it's taking the calculator a long time to evaluate an entry in a user-defined table and this inspires you to terminate the process by pressing [ALPHA], be warned — you will then have to clear the table and start over from scratch. Here's why: After pressing [ALPHA], you are confronted with an error message that asks you if you want to **Quit** or **Goto**. It would be nice if **Goto** sent you back to the table so you could edit your entry, but it doesn't; instead, it sends you to the definition of the sequence. So no matter which choice you make (**Quit** or **Goto**), the only way back to the table is by pressing 2nd GRAPH. But the entry that caused the problem in the first place is still in the table. So the calculator starts evaluating it all over again before displaying the table. This is indeed a vicious cycle! The only way out is to clear the table (as explained in the next section) and start over.

If the table displays the word ERROR instead of a sequence value, then the calculator is unable to determine what that value should be. The sequence may be undefined or not a real number. This can also happen when the value of **TblStart** in the Table Setup editor is less than the value of ***n*Min** in the Window editor. See Chapter 13 for another explanation of why this can happen.

While displaying the table for a sequence, you can edit the definition of the sequence without going back to the Y= editor. To do so, use the ► ◄ ▲ ▼ keys to place the cursor on the column heading for that sequence and then press ENTER. Edit the definition of the sequence and press ENTER when you're finished. The calculator automatically updates the table and the definition of the sequence in the Y= editor.

Clearing a User-Defined Table

After you have created a user-defined table (as described in the previous section), follow these steps to clear its contents:

1. **Press 2nd WINDOW to access the Table Setup editor and then set Indpend to Auto.**

2. **Press 2nd GRAPH to display an automatically generated table.**

3. **Press 2nd WINDOW and set Indpend back to Ask.**

4. **Press 2nd GRAPH to display an empty table.**

An automatically generated table cannot be cleared. To change the contents of such a table, change the values assigned to **TblStart** and Δ**Tbl** in the Table Setup editor. Automatically generated tables and the Table Setup editor are described in the previous section.

Viewing the Table and the Graph on the Same Screen

After you've graphed your sequences (as described in Chapter 13) and created a table for the sequences (as described earlier in this chapter), you can view the graph and the table on the same screen. To do so, follow these steps:

1. **Press** MODE.

2. **Put the calculator in G-T screen mode.**

   To do so, use the ▶◀▲▼ keys to place the cursor on **G-T** in the lower-right corner of the Mode menu and press ENTER to highlight it. This is illustrated in Figure 14-5.

Figure 14-5: Setting the mode for viewing a graph and a table.

3. **Press** GRAPH.

   After you press GRAPH, the graph and the table appear on the same screen (as illustrated in Figure 14-6).

If you press any key used in graphing a sequence, such as ZOOM or TRACE, the cursor will become active on the graph side of the screen. To return the cursor to the table, press 2nd GRAPH.

If you press TRACE and then use the ▶◀▲▼ keys to trace the graph, the value of the independent variable n corresponding to the cursor location on the graph will be highlighted in the table and the column for the sequence you're tracing will appear next to it. If necessary, the calculator updates the table so you can see that row in the table (as illustrated in the second picture in Figure 14-6).

To view the graph or the table in Full Screen mode, follow these steps:

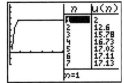 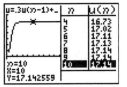

Figure 14-6: Graph-Table split screens.

1. **Press** MODE.

2. **Put the calculator in Full Screen mode.**

    To do so, use the ▶◀▲▼ keys to place the cursor on **Full** in the bottom left corner of the Mode menu and press ENTER to highlight it.

3. **Press** GRAPH **to see the graph, or press** 2nd GRAPH **to see the table.**

Evaluating Sequences

To evaluate a sequence at a specified value of *n*, follow these steps:

1. **Graph the sequences in a viewing window that contains the specified value of *n*.**

    Graphing sequences and setting the viewing window are explained in Chapter 13. To get a viewing window containing the specified value of *n*, that value must be between *n*Min and *n*Max.

2. **Set the Format menu to ExprOn and CoordOn.**

    Setting the Format menu is explained in Chapter 13.

3. **Press** 2nd TRACE **to access the Calculate menu.**

4. **Press** 1 **to select the value option.**

5. **Enter the specified value of *n*.**

    To do so, use the keypad to enter the value of *n*. This is illustrated in the first graph in Figure 14-7. (The graph in this figure is graphed in the Time format.) If you make a mistake when entering your number, press CLEAR and re-enter the number.

| Enter *n* | Press ENTER | Press ▲ |

Figure 14-7: Steps in evaluating sequences at a specified value of *n*.

6. **Press** ENTER.

What you see on the screen depends on the format used to produce the graph. The formats are as follows:

- **Time format:** After you press ENTER, the first highlighted sequence in the Y= editor appears at the top of the screen, the cursor appears on the graph of that function at the specified value of *n*, and the coordinates of the cursor appears at the bottom of the screen (as in the second graph in Figure 14-7).

   Repeatedly press ▲ to see the value of the other graphed sequences at your specified value of T. Each time you press ▲ the name of the sequence being evaluated appears at the top of the screen and the coordinates of the cursor location appear at the bottom of the screen (as in the third graph in Figure 14-7).

- **Web format:** In this format, the specified value of *n* and the coordinates of the cursor appear at the bottom of the screen. The *y*-coordinate is the value of the sequence at the specified value of *n*.

- **uv, vw, and uw format:** In this format, the specified value of *n* and values for *x* and *y* appear at the bottom of the screen. If you are in **uv** format, *x* is the value of $u(n)$ and *y* is the value of $v(n)$. In **vw** format, $x = v(n)$ and $y = w(n)$, and so on.

After using the **value** command to evaluate your sequence at one value of *n*, you can evaluate your sequence at another value of *n* by simply keying in the new value of *n* and then pressing ENTER. Pressing any function key (such as ENTER or TRACE) *after* evaluating a function will deactivate the **value** command.

If you're planning on evaluating sequences at several specified values of *n*, consider constructing a user-defined table.

Chapter 15

Parametric Equations

- -

In This Chapter

▶ Entering and graphing parametric equations

▶ Saving parametric graphs for future use

▶ Drawing on a parametric graph

▶ Tracing the graph of a pair of parametric equations

▶ Constructing a table of values of parametric equations

▶ Viewing the graph and the table on the same screen

▶ Evaluating parametric equations

- -

*A*s a particle moves along a curve, you usually want to know two things about that particle: Where is it? How long did it take to get there?

An answer to the first question is pretty straightforward: the *x*- and *y*-coordinates of the particle. To get an answer to the second question, you can bring a third variable — time — into the picture. To answer both questions simultaneously, you express the *x* and *y* variables as functions of *t*, the time variable.

The motion of the particle is thus described by the pair of parametric equations $x = f(t)$ and $y = g(t)$. In these equations, *t* is called the *parameter*, hence the equation type: *parametric*.

In addition to motion, such equations also describe graphs that are not the graphs of functions. An example of such a graph appears in Figure 15-5.

Entering Parametric Equations

The calculator can handle up to six pairs of parametric equations, X_{1T} and Y_{1T} through X_{6T} and Y_{6T}. To enter parametric equations in the calculator:

1. **Press** MODE **and put the calculator in Parametric mode, as shown in Figure 15-1.**

   To highlight an item in the Mode menu, use the ▶◀▲▼ keys to place the cursor on the item, and then press ENTER. Highlight **Par** in the fourth line to put the calculator in Parametric mode. For more information on the other items on the Mode menu please refer to Chapter 1.

Figure 15-1: Setting Parametric mode.

2. **Press** Y= **to access the Y= editor.**

3. **Enter the definitions of the pair of parametric equations** X_{nT} **and** Y_{nT}, **with n being an integer between 1 and 6.**

   To erase an entry that appears after X_{nT} or Y_{nT}, use the ▶◀▲▼ keys to place the cursor to the right of the equal sign and press CLEAR. Then enter your definition for the new parametric equation.

   When defining parametric equations, the only symbol the calculator allows for the independent variable (parameter) is the letter **T**. Press X,T,Θ,*n* to enter this letter in the definition of your parametric equation. In Figure 15-2, this key was used to enter the X_{1T} parametric equation.

   As a time saving convenience, when entering a parametric equation in the Y= editor, you can reference another parametric equation in its definition. In Figure 15-2, equations Y_{1T}, X_{3T}, and Y_{3T} are defined in this manner. To paste a parametric equation name in the equation you're entering in the Y= editor, press VARS▶2 and then press the number key for the name of the parametric equation you want pasted in the definition.

   Parametric equations must be entered in pairs. If you enter just one equation in the pair, say X_{nT} but not Y_{nT}, then when you go to graph these parametric equations, you won't see a graph nor will you get an error message warning you that there is a problem.

Figure 15-2: Examples of entering parametric equations.

Graphing Parametric Equations

After you have entered the parametric equations into the calculator, you can use the following steps to graph the equations:

1. **Turn off any Stat Plots that you don't want to appear in the graph of your parametric equations.**

   The first line in the Y= editor tells you the graphing status of the Stat Plots. (Stat Plots are discussed in Chapter 19.)

   If **Plot1**, **Plot2**, or **Plot3** is highlighted, then that Stat Plot will be graphed along with the graph of your parametric equations. If it's not highlighted, it won't be graphed. In Figure 15-2, **Plot1** will be graphed along with the parametric equations.

   To turn off a highlighted Stat Plot in the Y= editor, use the ▶◀▲▼ keys to place the cursor on the highlighted Stat Plot and then press ENTER. The same process is used to rehighlight the Stat Plot in order to graph it at a later time.

   When graphing parametric equations, Stat Plots that are turned on when you don't really want them to be graphed cause problems. The most common problem is the ERR: INVALID DIM error message. This error message gives you little insight into what is causing the problem. So if you aren't planning to graph a Stat Plot along with your parametric equations, please make sure all Stat Plots are turned off.

2. **Press** 2nd ZOOM **to access the Format menu.**

3. **Set the format for the graph by using the ▶◀▲▼ keys to place the cursor on the desired format and then press** ENTER.

   Figure 15-3 pictures the Format menu when the calculator is in Parametric mode. The first line of the menu gives you a choice between having points on the graph displayed in (x, y) rectangular form or in (r, θ) polar form. The remaining items on the Format menu are explained in Chapter 9. When graphing parametric equations, there is no harm in leaving these items highlighted as they appear in Figure 15-3.

Figure 15-3: Format menu in Parametric mode.

4. **Press** WINDOW **to access the Window editor.**

5. **After each of the first three window variables, enter a numerical value that is appropriate for the parametric equations you're graphing. Press** ENTER **after entering each number.**

Figure 15-4 pictures the Window editor when the calculator is in Parametric mode. The following gives an explanation of the variables that you must set in this editor:

• **Tmin:** This setting contains the first value of the independent variable (parameter) **T** that the calculator will use to evaluate all parametric equations in the Y= editor. It can be set equal to any real number.

• **Tmax:** This setting contains the largest value of the independent variable (parameter) **T** that you want the calculator to use to evaluate all parametric equations in the Y= editor. It can be set equal to any real number.

• **Tstep:** This setting tells the calculator how to increment the independent variable **T** as it evaluates the parametric equations in the Y= editor and graphs the corresponding points.

Tstep must be a positive real number.

You want **Tstep** to be a small number like 0.1 or $\pi/24$, but you don't want it to be too small like 0.001. If **Tstep** is too small, it will take a few minutes for the calculator to produce the graph. And if **Tstep** is a large number like 1, the calculator may not graph enough points to show you the true shape of the curve.

If it's taking a long time for the calculator to graph your parametric equations, and this causes you to regret that very small number you placed in **Tstep,** press ON to terminate the graphing process. You can then go back to the Window editor and adjust the **Tstep** setting.

When graphing parametric equations that use trigonometric functions in their definition, it's best to set **Tmin, Tmax,** and **Tstep** to multiples of π. In Figure 15-4, **Tmax** is set equal to 2π and **Tstep** is set equal to $\pi/24$. π is entered into the calculator by pressing 2nd ^.

```
WINDOW
 Tmin=0
 Tmax=6.2831853…
 Tstep=.1308996…
 Xmin=-10
 Xmax=10
 Xscl=1
↓Ymin=-10
```

Figure 15-4: Window editor in Parametric mode.

The remaining items in the Window editor deal with setting a viewing window for the graph. Setting the viewing window is explained in detail in Chapter 9. If you know the dimensions of the viewing window required for your graph, go ahead and assign numerical values to the remaining variables in the Window editor and advance to Step 8. If you don't know the minimum and maximum x and y values required for you graph, the next step tells you how to get the calculator to find them for you.

6. **Press** ZOOM 0 **to access ZoomFit.**

After you've assigned values to **Tmin, Tmax,** and **Tstep,** ZoomFit determines appropriate values for **Xmin, Xmax, Ymin,** and **Ymax,** and then graphs your parametric equations. Note, however, that **ZoomFit** graphs parametric equations in the smallest possible viewing window, and it doesn't assign new values to **Xscl** and **Yscl.**

If you like the way the graph looks, then you can skip the remaining steps. If you'd like spruce up the graph, Step 7 gives you some pointers.

7. **Press** WINDOW **and adjust the values assigned to the X and Y settings. Press** ENTER **after entering each new number.**

Here are some pointers on how to readjust the viewing window after using **ZoomFit:**

- **Xmin and Xmax:** If you don't want the graph to hit the left and right sides of the screen, decrease the value assigned to **Xmin** and increase the value assigned to **Xmax.** If you want to see the *y*-axis on the graph, assign values to **Xmin** and **Xmax** so that zero is strictly between these two values.

- **Xscl:** Set this equal to a value that doesn't make the *x*-axis look like railroad tracks — that is, an axis with far too many tick marks. Twenty or fewer tick marks makes for a nice looking axis.

- **Ymin and Ymax:** If you don't want the graph to hit the top and bottom of the screen, decrease the value assigned to **Ymin** and increase the value assigned to **Ymax.** If you want to see the *x*-axis on the graph, assign values to **Ymin** and **Ymax** so that zero is strictly between these two values.

- **Yscl:** Set this equal to a value that doesn't make the *y*-axis look like railroad tracks. Fifteen or fewer tick marks is a nice number for the *y*-axis.

8. **Press** GRAPH **to regraph the parametric equations.**

After you've graphed your parametric equations (as described earlier in this chapter), you can draw lines and functions on the graph or write text on the graph. And you can save a picture of the graph and the drawings.

Chapter 12 describes the procedures for drawing on a graph and for saving the result as a picture.

Graphing several equations

To identify the graphs of several pairs of parametric equations, set a different graph style for each pair of parametric equations. The following steps show how:

1. **Press** Y= **to access the Y= editor.**

2. **Use the** ▶◀▲▼ **keys to place the cursor on the icon appearing at the very left of the definition of the first parametric equation in the pair of parametric equations.**

3. **Repeatedly press** ENTER **until you get the desired graph style.**

You have five styles to choose from: Line, Thick Line, Path, Animate, and Dotted Line. Each time you press ENTER, you get a different style. The Path style, denoted by the ⊹ icon, uses a circle to indicate a point as it's being graphed, and when the graph is complete, the circle disappears and leaves the graph in Line style. The Animate style, denoted by the ◊ icon, also uses a circle to indicate a point as it's being graphed, but when the graph is complete, no graph appears on-screen.

The most common styles used to graph parametric equations are Line, Thick Line, and Dotted Line. In Figure 15-2, X_{1T} and Y_{1T} are set to the default Line style; X_{2T} and Y_{2T} are set to the Thick Line style; X_{3T} and Y_{3T} are set to the Dotted Line style.

If you don't want the calculator to graph a pair of parametric equations, unhighlight the equal signs in that pair. To graph it at a later time, rehighlight the equal signs. You can do so in the Y= editor by using the ▶◀▲▼ keys to place the cursor on the equal sign in the definition of either parametric equation in the pair, and then pressing ENTER to toggle the equal sign between highlighted and un-highlighted. In Figure 15-2, for example, the calculator won't graph the parametric equations X_{1T} and Y_{1T}, but it will graph the other two pairs of parametric equations.

Panning in Parametric mode

When you're tracing parametric equations and the cursor hits an edge of the screen, if **T** is less than **Tmax** and you continue to press ▶, the coordinates of the cursor will be displayed at the bottom of the screen, but the calculator does not automatically adjust the viewing window. So you won't see the cursor on the graph. To rectify this situation, make a mental note of the value of **T** and then press ⬛. The calculator then redraws the graph centered at the location of the cursor at the time you pressed ENTER.

Unfortunately, after the graph is redrawn, the trace cursor is placed at the beginning of the *first* pair of parametric equations appearing in the Y= editor — and it may not even appear on-screen. To get the trace cursor back on the part of the graph displayed in the new viewing window, key in that value of **T** that you made a mental note of earlier, and then press ENTER. If you weren't tracing the first pair of parametric equations that appear in the Y= editor, use ▲ to place the cursor on the graph you were tracing. The trace cursor then appears in the middle of the screen and you can use ▶ to continue tracing the graph.

Using ZOOM commands

After you've graphed your parametric equations (as explained earlier in this chapter), you can use Zoom commands to change the viewing window of your graph. Pressing ZOOM accesses the Zoom menu. On this menu you will see all the Zoom commands that are available for graphing functions. Chapter 10 offers a detailed explanation of what these commands do and how you use them.

The Zoom commands that are most useful when graphing parametric equations are

✔ **ZoomFit:** This command finds a viewing window for a specified portion of the graph. How to use **ZoomFit** is explained in the section, "Graphing Parametric Equations," earlier in this chapter.

✔ **ZSquare:** Because the calculator screen is rectangular instead of square, circles will look like ellipses if the viewing window isn't properly set. **ZSquare** adjusts the settings in the Window editor so that circles look like circles. To use **ZSquare**, first graph your parametric equations and then press ZOOM 5. The graph is then redrawn in a viewing window that makes circles look like circles. Figure 15-5 illustrates the effect that **ZSquare** has on the spiral $x = t\cos(t)$, $y = t\sin(t)$.

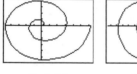

    ZoomFit   ZSquare

Figure 15-5: A spiral graphed using **ZoomFit** and then using **ZSquare**.

✔ **Zbox:** Use this command to define a new viewing window for a portion of your graph by enclosing it in a box (as shown in Figure 15-6). The box then becomes the new viewing window. To construct the box, follow these steps:

    1. Press ZOOM 1.

    2. Define a corner of the box.

        To do so, use the ▶ ◀ ▲ ▼ keys to move the cursor to the spot where you want one corner of the box to be located, and then press ENTER. The calculator marks that corner of the box with a small square.

3. Construct the box.

To do so, press the ▶◀▲▼ keys. As you press these keys, the calculator draws the sides of the box.

When you use **ZBox**, if you don't like the size of the box, you can use any of the ▶◀▲▼ keys to resize the box. If you don't like the location of the corner you anchored in Step 2, press CLEAR and start over with Step 1.

4. When you're finished drawing the box, press ENTER and the graph is redrawn in the viewing window specified by the box.

When you use **ZBox**, ENTER is pressed only two times. The first time you press it's to anchor a corner of the zoom box. The next time you press ENTER is when you're finished drawing the box and are ready to have the calculator redraw the graph.

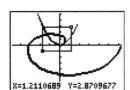

X=1.2110689 Y=2.8709677

Figure 15-6: Using ZBox.

✔ **Zoom In and Zoom Out:** After the graph is drawn, these commands allow you to zoom in on a portion of the graph or to zoom out from the graph. They work very much like a zoom lens on a camera. To use these commands, follow these steps:

1. Press ZOOM 2 if you want to zoom in, or press ZOOM 3 if you want to zoom out.

2. Use the ▶◀▲▼ keys to move the cursor to the spot on the screen from which you want to zoom in or zoom out.

This spot becomes the center of the new viewing window.

3. Press ENTER.

You can press ENTER again to zoom in closer or to zoom out one more time.

4. Press CLEAR when you're finished zooming in or zooming out.

If you used a Zoom command to redraw a graph and then want to undo what that command did to the graph, press $\boxed{\text{ZOOM}}\boxed{\blacktriangleright}\boxed{1}$ to select **ZPrevious** from the Zoom Memory menu. **ZPrevious** redraws the graph as it appeared in the previous viewing window.

Saving a parametric graph

After you have graphed your parametric equations, as described earlier in this chapter, you can save the graph and its entire Window, Y= editor, Mode, and Format settings in a Graph Database. When you recall the Database at a later time, you get more than just a picture of the graph. The calculator also restores the Window, Y= editor, Mode, and Format settings to those stored in the database. So you can, for example, trace the recalled graph.

The procedures for saving and recalling a Graph Database are described in Chapter 9.

Tracing a parametric graph

After you've graphed your parametric equations (as described earlier in this chapter), press $\boxed{\text{TRACE}}$ and then use $\boxed{\blacktriangleright}$ to investigate these equations. Here's what you see, and what you can do to change things:

- ✔ **The definition of the parametric equations:** The parametric equations you're tracing are displayed at the top of the screen, provided the calculator is in **ExprOn** format, as indicated in Figure 15-3. If you've graphed more than one pair of parametric equations and you'd like to trace a different pair of equations, press $\boxed{\blacktriangle}$. Each time you press this key, the cursor jumps to another pair of parametric equations. Eventually it jumps back to the original pair of equations.

- ✔ **The independent variable T:** The value of **T** corresponding to the cursor location is displayed in the lower left hand corner of the screen, provided the calculator is in **CoordOn** format, as indicated in Figure 15-3. When you press $\boxed{\text{TRACE}}$, the cursor is placed at the beginning of the graph of the parametric equations and **T** displays the value you assigned to **Tmin** in the Window editor. Each time you press $\boxed{\blacktriangleright}$, the cursor moves to the next plotted point in the graph and the value of **T** changes to the value of the independent variable corresponding to that plotted point.

  If you press $\boxed{\blacktriangleleft}$, the cursor will move left to the previously plotted point in the graph. And if you press $\boxed{\blacktriangle}$ to trace a different pair of parametric equations, the calculator starts tracing that pair of equations at the value of **T** that was displayed on-screen before you pressed this key.

✔ **The values of *x* and *y*:** At the bottom of the screen, you see the values of the *x*- and *y*- coordinates of the cursor location (provided the calculator is in **CoordOn** format), as in Figure 15-3. In the **PolarGC** format, the coordinates of this point display in polar form.

Press CLEAR to stop tracing the graph. Doing so also removes the name of the function and the coordinates of the cursor from the screen.

If your cursor disappears from the screen while you are tracing a parametric graph, see the sidebar "Panning in Parametric mode" to see how to rectify this situation.

When you're using TRACE, if you want to start tracing your parametric equations at a specific value of the independent variable **T**, just key in that value and press ENTER. (The value you assign to **T** must be between **Tmin** and **Tmax**; if it's not, you will get an error message.) After you press ENTER, the trace cursor moves to the point on the graph corresponding to that value of **T**. If that point isn't on the part of the graph that's on-screen, you can get the cursor and the graph to appear in the same viewing window. The sidebar, "Panning in Parametric mode," tells you how to do so.

If the name of the parametric equations and the value of the independent variable **T** are interfering with your view of the graph when you use TRACE, press WINDOW and decrease the value of **Ymin** and increase the value of **Ymax**.

Displaying Equations in a Table

After you have entered the parametric equations into the calculator, as described earlier in this chapter, you can have the calculator create a table of the terms in all pairs of parametric equations in the Y= editor that are defined with a highlighted equal sign. (The tip in the "Graphing several equations" section of this chapter tells you how to highlight or unhighlight the equal sign.) To create a table:

1. **Highlight the equal sign of those parametric equations in the Y= editor that you want to appear in the table.**

   Only those parametric equations in the Y= editor that are defined with a highlighted equal sign will appear in the table. To highlight or unhighlight an equal sign, press Y=, use the ▶◀▲▼ keys to place the cursor on the equal sign in the definition of either equation in the pair of parametric equations, and then press ENTER to toggle the equal sign between highlighted and unhighlighted.

2. **Press 2nd WINDOW to access the Table Setup editor (shown in Figure 15-7).**

```
TABLE SETUP
 TblStart=0
  ΔTbl=.1
Indpnt: Auto Ask
Depend: Auto Ask
```

Figure 15-7: The Table Setup editor.

3. **Enter a number in TblStart and then press ENTER.**

   TblStart is the first value of the independent variable **T** to appear in the table. In Figure 15-7, **TblStart** is assigned the value 0.

   To enter the number you have chosen for **TblStart**, place the cursor on the number appearing after the equal sign, press the number keys to enter your number, and then press ENTER.

4. **Enter a number in ΔTbl and then press ENTER.**

   ΔTbl gives the increment for the independent variable **T**. You usually want to set **ΔTbl** equal to a small number like 0.5, 0.1, or 0.01. In Figure 15-7, **ΔTbl** is assigned the value 0.1.

   To enter the number you have chosen for **ΔTbl**, place the cursor on the number appearing after the equal sign, press the number keys to enter your number, and then press ENTER.

5. **Set the mode for Indpnt and Depend.**

   To change the mode of either **Indpnt** or **Depend**, use the ▶◀▲▼ keys to place the cursor on the desired mode, either **Auto** or **Ask**, and then press ENTER.

   To have the calculator automatically generate the table for you, put both **Indpnt** and **Depend** in **Auto** mode. The first table in Figure 15-8 was constructed in this fashion.

   If you want to create a user-defined table that contains only those values of the independent variable **T** that you

specify — and then have the calculator figure out the corresponding values of the parametric equations — put **Indpnt**
in **Ask** mode and **Depend** in **Auto** mode. Step 6 explains how
you construct such a table; the second table in Figure 15-8,
constructed in this fashion, shows you how it looks.

For a user-defined table, you don't have to assign values to
TblStart and Δ**Tbl** in the Table Setup editor.

The other combinations of mode settings for **Indpnt** and
Depend are not that useful when you're dealing with parametric equations.

T	X₁ᴛ	Y₁ᴛ
0	6	0
.1	5.995	.5995
.2	5.98	1.196
.3	5.955	1.7865
.4	5.92	2.368
.5	5.875	2.9375
.6	5.82	3.492
T=0		

T	X₁ᴛ	Y₁ᴛ
0	6	0
1.5	4.875	7.3125
3	1.5	4.5
3.7	-.845	-3.127
3.1416	1.0652	3.3464
T=2π		

Automatically   User-defined
generated

Figure 15-8: Tables for parametric equations.

6. **Press** ⟨2nd⟩⟨GRAPH⟩ **to display the table.**

   When you display the table, what you see on the screen
   depends on the modes you set for **Indpnt** and **Depend** in
   Step 5. And what you can do with the table also depends on
   these modes. Here's what you see and what you can do:

   - **For an automatically generated table:**

     If **Indpnt** and **Depend** are both in **Auto** mode, then
     pressing ⟨2nd⟩⟨GRAPH⟩ generates the table automatically.
     To display rows in the table beyond the last row on the
     screen, repeatedly press ⟨▾⟩ until those rows appear.
     Likewise, repeatedly press ⟨▴⟩ to display rows above
     the first row on the screen.

     Each time the calculator redisplays a table with a differ-
     ent set of rows, it also automatically resets **TblStart** to
     the value of **T** appearing in the first row of the newly
     displayed table. To return the table to its original state,
     press ⟨2nd⟩⟨WINDOW⟩ to access the Table Setup editor, and
     then change the value the calculator assigned to **TblStart**.

     If you're constructing a table for more than one pair of
     parametric equations, only the first pair will appear on
     the screen. To see another pair of equations, repeatedly
     press ⟨▸⟩ until they appear. This causes the first pair of
     equations to disappear. To see them again, repeatedly
     press ⟨◂⟩ until they appear.

- **For a user-generated table:**

  If you put **Indpnt** in **Ask** mode and **Depend** in **Auto** mode so that you can generate your own table, then the table should be empty when you display it. If it's not, clear the table (as described in the next section).

  In an empty table, key in the first value of the independent variable **T** you want to appear in the table, and then press ENTER. The corresponding values of the first pair of parametric equations automatically appear. Key in the next value of **T** you want in the table and press ENTER, and so on. The values for **T** that you place in the first column don't have to be in any specific order, nor do they have to be between **Tmin** and **Tmax**.

While displaying the table for parametric equations, you can edit the definition of the equations without going back to the Y= editor. To do so, use the ▶◀▲▼ keys to place the cursor on the column heading for that parametric equation and then press ENTER. Edit the definition of the equation (editing expressions is explained in Chapter 1) and press ENTER when you're finished. The calculator automatically updates the table and the definition of the parametric equation in the Y= editor.

The word ERROR appearing in a table indicates that either the parametric equations are undefined or the corresponding values of **T** are not real numbers.

Clearing a user-defined table

After you have created a user-defined table, as described in the previous section, to clear its contents, perform the following steps:

1. **Press** 2nd WINDOW **to access theTable Setup editor and then set Indpend to Auto.**

2. **Press** 2nd GRAPH **to display an automatically generated table.**

3. **Press** 2nd WINDOW **and set Indpend back to Ask.**

4. **Press** 2nd GRAPH **to display an empty table.**

An automatically generated table cannot be cleared. To change the contents of such a table, change the values assigned to **TblStart** and **ΔTbl** in the Table Setup editor.

Viewing the table and the graph on the same screen

After you have graphed your parametric equations and created a table for the parametric equations, you can view the graph and the table on the same screen. To do so, follow these steps:

1. Press MODE.

2. Put the calculator in G-T screen mode.

To do so, use the ▶◀▲▼ keys to place the cursor on **G-T** in the bottom-right corner of the Mode menu and press ENTER to highlight it. This is illustrated in Figure 15-9.

Figure 15-9: Setting the mode for viewing a graph and a table.

3. Press GRAPH.

After you press GRAPH, the graph and the table appear on the same screen.

If you press any key used in graphing parametric equations, such as ZOOM or TRACE, the cursor will become active on the graph side of the screen. To return the cursor to the table, press 2nd GRAPH.

If you press TRACE and then use the ▶◀▲▼ keys to trace the graph, the value of the independent variable **T** corresponding to the cursor location on the graph will be highlighted in the table and the columns for the parametric equations you're tracing will appear next to it. If necessary, the calculator will update the table so you can see that row in the table. This is illustrated in Figure 15-10.

To view the graph or the table in full screen mode:

1. Press MODE.

2. Put the calculator in Full screen mode.

To do so, use the ▶◀▲▼ keys to place the cursor on **Full** in the bottom left hand corner of the Mode menu and press ENTER to highlight it.

3. Press GRAPH to see the graph, or press 2nd GRAPH to see the table.

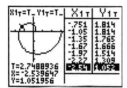

Figure 15-10: Using TRACE in Graph-Table mode.

Evaluating Parametric Equations

To evaluate a pair of parametric equations at a specified value of **T**:

1. Graph the equations in a viewing window that contains the specified value of T.

Graphing parametric equations and setting the viewing window are explained earlier in this chapter. To get a viewing window containing the specified value of **T**, that value must be between **Tmin** and **Tmax**.

2. Set the Format menu to ExprOn and CoordOn.

Setting the Format menu is explained earlier in this chapter.

3. Press 2nd TRACE to access the Calculate menu.

4. Press 1 to select the value option.

5. Enter the specified value of T.

To do so, use the keypad to enter the value of **T**. This is illustrated in the first graph in Figure 15-11. If you make a mistake when entering your number, press CLEAR and re-enter the number.

6. Press ENTER.

After you press ENTER, the first highlighted pair of parametric equations in the Y= editor appear at the top of the screen, the cursor appears on the graph of that function at the specified value of **T**, and the coordinates of the cursor appear at the bottom of the screen. This is illustrated in the second graph in Figure 15-11.

7. **Repeatedly press ▲ to see the value of the other graphed parametric equations at your specified value of T.**

   Each time you press ▲ the names of the parametric equations being evaluated appear at the top of the screen; the coordinates of the cursor location appear at the bottom of the screen. The third graph in Figure 15-11 shows what this arrangement looks like.

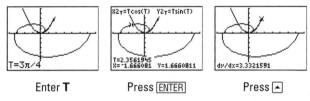

Enter **T**	Press ENTER	Press ▲

Figure 15-11: Steps in evaluating equations at a specified value of **T**.

If you're planning on evaluating parametric equations at several specified values of **T**, consider constructing a user-defined table.

Finding Derivatives

To find the derivative (*dy/dx*, *dy/dt*, or *dx/dt*) of a pair of parametric equations at a specified value of **T**:

1. **Graph the parametric equations in a viewing window that contains the specified value of T.**

   Graphing parametric equations and setting the viewing window are explained earlier in this chapter. If you want to get a viewing window that contains the specified value of **T**, make sure that value is between **Tmin** and **Tmax**.

2. **Set the Format menu to ExprOn and CoordOn.**

   Setting the Format menu is explained earlier in this chapter.

3. **Press 2nd TRACE to access the Calculate menu.**

4. **Press 2, 3, or 4 to respectively select the dy/dx, dy/dt, or dx/dt option.**

5. **If necessary, repeatedly press ⌃ until the appropriate parametric equations appear at the top of the screen.**

   This is illustrated in the first graph in Figure 15-12.

6. **Enter the specified value of T.**

   To do so, use the keypad to enter the value of **T**. As you use the keypad, **T=** replaces the coordinates of the cursor location appearing at the bottom of the screen in the previous step. The number you key in is placed after **T=** (as in the second graph in Figure 15-12). If you make a mistake when entering your number, press [CLEAR] and re-enter the number.

   If you're interested only in finding the slope of the curve in a general area of the graph (instead of at a specific value of **T**), then instead of entering a value of **T**, just use the ◂ and ▸ to move the cursor to the desired location on the graph.

7. **Press [ENTER].**

   After pressing [ENTER], the derivative is displayed at the bottom of the screen. This is illustrated in the third graph in Figure 15-12.

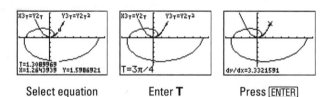

   Select equation   Enter **T**   Press [ENTER]

Figure 15-12: Steps in finding the *dy/dx* at a specified value of T.

Chapter 16

Polar Equations

• •

In This Chapter

▶ Converting between polar and rectangular coordinates

▶ Entering and graphing polar equations

▶ Tracing the graph of a polar equation

▶ Constructing a table of values of polar equations

▶ Evaluating polar equations at a specified value of θ

▶ Finding derivatives at specified values of θ

• •

*I*n the polar coordinate system, a point in the plane is described by its angular direction (θ) and its distance *r* from the origin. The angular direction θ is always measured from the positive *x*-axis and the polar coordinates of the point are denoted by (*r*, θ). This is illustrated in Figure 16-1.

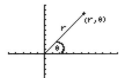

Figure 16-1: Representing a point in polar coordinates.

Converting Coordinates

We're used to seeing points in the *xy*-plane displayed in the (*x*, *y*) rectangular form. But when dealing with polar equations, we view these points in their (*r*, θ) polar form. To convert (*x*, *y*) from rectangular to polar form:

    1. **Press** 2nd APPS **to access the Angle menu. (On the TI-83, press** 2nd MATRX **.)**

2. **Press ⑤ to select the** R ▸ Pr(**option.**

   This is illustrated in the first picture in Figure 16-2.

3. **Enter the point (x, y) and press** ENTER **to display r in the (r, θ) polar form of the point.**

   To do so, enter the x-coordinate, press ⦋,⦌, enter the y-coordinate, press ⦋)⦌, and then press ENTER. The value of r appears on the right of the screen. This is illustrated in the second picture in Figure 16-2.

4. **Press** 2nd APPS ⑥ **to select the** R ▸ PΘ(**option from the Angle menu. (On the TI-83, press** 2nd MATRX ⑥**.)**

5. **Enter the point (x, y) and press** ENTER **to display θ in the (r, θ) polar form of the point.**

   This is illustrated in the second picture in Figure 16-2. θ will be in degrees if the calculator is in Degree mode, or it will be measured in radians if the calculator is in Radian mode.

Angle menu   Polar form

Figure 16-2: Converting from rectangular to polar coordinates.

If you're a purist like me, you like radian measures, such as the one in Figure 16-2, to be expressed as a fractional multiple of π, if possible. It's much easier to know what the angle looks like in this form. I can quickly sketch the angle π/6, but it would take me awhile to sketch the angle .5235987756 radians. It turns out that they are the same angle. This is why I like to convert a radian measure to a fractional multiple of π, using the following steps:

1. **If the radian measure is not the last entry in the calculator, enter it.**

   In the second picture in Figure 16-2, the last line indicates that the radian measure of θ is the last entry in the calculator.

2. **Press** ÷ 2nd ^ ENTER **to divide the radian measure by π.**

3. **Press** MATH ① ENTER **to display the fractional multiple of π.**

If the calculator can't convert the decimal to a fraction, it just redisplays the decimal.

Figure 16-3 illustrates the above steps. The 1/6 in this figure indicates that the decimal radian measure (.5235987756) is equal to $\pi/6$, thus showing that $(2, \pi/6)$ is the polar form of the point $(\sqrt{3}, 1)$.

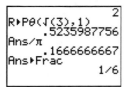

```
                     2
R►Pθ(√(3),1)
         .5235987756
Ans/π
          .1666666667
Ans►Frac
                  1/6
```

Figure 16-3: Converting a decimal to a fraction.

To convert (r, θ) from polar to rectangular coordinates:

1. **Press** 2nd APPS 7 **to select** R ► Px(**from the Angle menu. (On the TI-83, press** 2nd MATRX 7 **.)**

2. **Enter** *(r, θ)* **and press** ENTER **to display the *x*-coordinate of the point *(x, y)*.**

   To do so, enter *r*, press ․ , enter θ, press) , and then press ENTER. The value of *x*-coordinate appears on the right of the screen. If your calculator is in Degree mode, θ should be entered in degrees; if the calculator is in Radian mode, θ should be entered in radian measure.

   If the calculator is in Degree mode, but you want to enter an angle in radian measure, enter the radian measure of the angle (surrounded by parentheses, if necessary) and then press 2nd APPS 3 to indicate that it's in radian measure. (On the TI-83, press 2nd MATRX 3 .) If you want to indicate that an angle is in degree measure when the calculator is in Radian mode, press 2nd APPS 1 . (On the TI-83, press 2nd MATRX 1 .) These are illustrated in Figure 16-4.

```
P►Rx(2,(π/6)ʳ)
          1.732050808
P►Rx(2,30°)
          1.732050808
```

Figure 16-4: Indicating the measure, radian, or degree of an angle.

3. **Press** [2nd][APPS][8] **to select the** R ▸ Py(**option from the Angle menu. (On the TI-83, press** [2nd][MATRX][8].)

4. **Enter** (r, θ) **and press** [ENTER] **to display the y-coordinate of the point** (x, y).

Entering Polar Equations

The calculator can handle up to six polar equations, r_1 through r_6. To enter polar equations in the calculator, follow these steps:

1. **Press** [MODE] **and put the calculator in Polar mode, as shown in Figure 16-5.**

   To highlight an item in the Mode menu, use the [▸][◂][▲][▼] keys to place the cursor on the item, and then press [ENTER]. Highlight **Pol** in the fourth line to put the calculator in Polar mode.

Figure 16-5: Setting Polar mode.

2. **Press** [Y=] **to access the Y= editor.**

3. **Enter the definition of the polar equation** r_n, **n being an integer between 1 and 6.**

   To erase an entry that appears after r_n, use the [▸][◂][▲][▼] keys to place the cursor to the right of the equal sign and press [CLEAR]. Then enter your definition for the new polar equation.

When you're defining polar equations, the only symbol the calculator allows for the independent variable is the angle θ. Press [X,T,Θ,n] to enter θ in the definition of your polar equation. In Figure 16-6, this key was used to enter equations r_1, r_2, and r_3.

When you enter a polar equation in the Y= editor, as a timesaver you can reference another polar equation in the definition you're entering. In Figure 16-6, for example, equation r_4 is defined in this manner. To paste a polar equation name into the equation you're entering in the Y= editor, press [VARS][▸][3] and then press the number key for the name of the polar equation you want to paste into the definition.

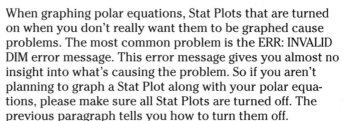

Figure 16-6: Examples of entering polar equations.

Graphing Polar Equations

After you've entered the polar equations into the calculator, as described in the previous section, you can use the following steps to graph the equations:

1. **Turn off any Stat Plots that you don't want to appear in the graph of your polar equations.**

   The first line in the Y= editor tells you the graphing status of the Stat Plots. (Stat Plots are discussed in Chapter 19.)

   If **Plot1, Plot2,** or **Plot3** is highlighted, then that Stat Plot will be graphed along with the graph of your polar equations. If it's not highlighted, it won't be graphed. In Figure 16-6, **Plot1** will be graphed along with the polar equations.

   To turn off a highlighted Stat Plot in the Y= editor, use the ▶◀▲▼ keys to place the cursor on the highlighted Stat Plot and then press [ENTER]. The same process is used to rehighlight the Stat Plot in order to graph it at a later time.

   When graphing polar equations, Stat Plots that are turned on when you don't really want them to be graphed cause problems. The most common problem is the ERR: INVALID DIM error message. This error message gives you almost no insight into what's causing the problem. So if you aren't planning to graph a Stat Plot along with your polar equations, please make sure all Stat Plots are turned off. The previous paragraph tells you how to turn them off.

2. **Press [2nd][ZOOM] to access the Format menu.**

3. **Set the format for the graph by using the ▶◀▲▼ keys to place the cursor on the desired format and then press [ENTER].**

   Figure 16-7 pictures the Format menu when the calculator is in Polar mode. The first line of the menu gives you a choice between having points on the graph displayed in (x, y) rectangular form or in (r, θ) polar form. The remaining items on the Format menu are explained in Chapter 9. When graphing polar equations, there is no harm in leaving these remaining items highlighted as they appear in Figure 16-7.

Figure 16-7: Format menu in Polar mode.

4. **Press** WINDOW **to access the Window editor.**

5. **After each of the first three window variables, enter a numerical value that is appropriate for the polar equations you're graphing. Press** ENTER **after entering each number.**

   Figure 16-8 shows the Window editor when the calculator is in Polar mode. The following gives an explanation of the variables that you must set in this editor:

   • **θmin:** This setting contains the first value of the independent variable θ that the calculator will use to evaluate all polar equations in the Y= editor. It can be set equal to any real number.

```
WINDOW
 θmin=0
 θmax=6.2831853…
 θstep=.1308996…
 Xmin=-10
 Xmax=10
 Xscl=1
↓Ymin=-10
```

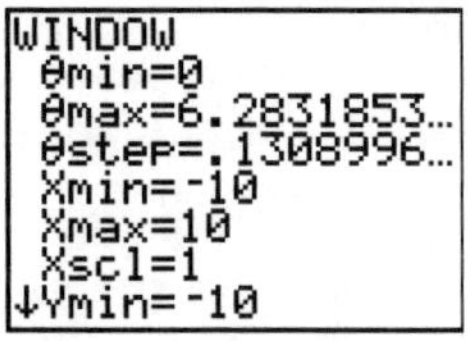

Figure 16-8: Window editor in Polar mode.

   • **θmax:** This setting contains the largest value of the independent variable θ that you want the calculator to use to evaluate all polar equations in the Y= editor. It can be set equal to any real number.

   θmax is usually set equal to 2π. However, as illustrated in Figure 16-9, there are occasions when you may want to set it equal to a larger number. π is entered into the calculator by pressing 2nd ^.

   • **θstep:** This setting tells the calculator how to increment the independent variable θ as it evaluates the polar equations in the Y= editor and graphs the corresponding points.

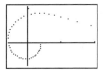

θmax = 2π   θmax = 4π

Figure 16-9: The graph of $r = 1/\sqrt{\Theta}$ with θmax = 2π and 4π.

θ**step** must be a positive real number.

You want θ**step** to be a small number like π/24, but you don't want it to be too small like 0.001. If θ**step** is too small, it will take a few minutes for the calculator to produce the graph. And if θ**step** is a large number like 1, the calculator may not graph enough points to show you the true shape of the curve.

If it's taking a long time for the calculator to graph your polar equations, and you start to regret that very small number you placed in θ**step**, press [ON] to terminate the graphing process. You can then go back to the Window editor and adjust the θ**step** setting.

The remaining items in the Window editor deal with setting a viewing window for the graph. Setting the viewing window is explained in detail in Chapter 9. If you know the dimensions of the viewing window required for your graph, go ahead and assign numerical values to the remaining variables in the Window editor and advance to Step 8. If you don't know the minimum and maximum *x* and *y* values required for your graph, the next step tells you how to get the calculator to find them for you.

6. **Press [ZOOM][0] to access ZoomFit.**

   After you've assigned values to θ**min,** θ**max,** and θ**step,** **ZoomFit** determines appropriate values for **Xmin, Xmax, Ymin,** and **Ymax** and graphs your polar equations. However, **ZoomFit** graphs the polar equations in the smallest possible viewing window and does not assign appropriate values to **Xscl** and **Yscl.**

   If you like the way the graph looks, then you can skip the remaining steps. If you'd like spruce up the graph, Step 7 offers some pointers.

7. **Press [WINDOW] and adjust the values assigned to the X and Y settings. Press [ENTER] after entering each new number.**

Here are some pointers on how to readjust the viewing window after using **ZoomFit:**

- **Xmin and Xmax:** If you don't want the graph to hit the left and right sides of the screen, decrease the value assigned to **Xmin** and increase the value assigned to **Xmax.** If you want to see the *y*-axis on the graph, assign values to **Xmin** and **Xmax** so that zero is strictly between these two values.

- **Xscl:** Set this equal to a value that doesn't make the *x*-axis look like railroad tracks — that is, an axis with far too many tick marks. Twenty or fewer tick marks makes for a nice looking axis.

- **Ymin and Ymax:** If you don't want the graph to hit the top and bottom of the screen, decrease the value assigned to **Ymin** and increase the value assigned to **Ymax.** If you want to see the *x*-axis on the graph, assign values to **Ymin** and **Ymax** so that zero is strictly between these two values.

- **Yscl:** Set this equal to a value that doesn't make the *y*-axis look like railroad tracks. Fifteen or fewer tick marks is a nice number for the *y*-axis.

8. **Press** GRAPH **to regraph the polar equations.**

After you've graphed your polar equations (as described earlier in this chapter), you can draw lines and functions on the graph. You can also write text on the graph. And you can save a picture of the graph and the drawings.

The procedures for drawing on a graph and for saving the result as a picture are described in Chapter 12.

Graphing several equations

To identify the graphs of several polar equations, set a different graph style for each polar equation:

1. **Press** Y= **to access the Y= editor.**

2. **Use the** ▶ ◀ ▲ ▼ **keys to place the cursor on the icon appearing at the very left of the definition of the polar equation.**

3. **Repeatedly press** ENTER **until you get the desired graph style.**

You have five styles to choose from: Line, Thick Line, Path, Animate, and Dotted Line. Each time you press ENTER, you get a different style. The Path style, denoted by the ⁴⁰ icon, uses a circle to indicate a point as it's being graphed, and when the graph is complete, the circle disappears and leaves the graph in Line style. The Animate style, denoted by the ⁴ icon, also uses a circle to indicate a point as it's being graphed, but when the graph is complete, no graph appears on the screen.

The most common styles used to graph polar equations are Line, Thick Line, and Dotted Line. In Figure 16-6, r_1 is set to the default Line style; r_2 is set to the Thick Line style; r_3 and r_4 are set to the Dotted Line style.

If you don't want the calculator to graph a polar equation, un-highlight the equal sign in that equation. To graph it at a later time, rehighlight the equal sign. To do so in the Y= editor, use the ▶◀▲▼ keys to place the cursor on the equal sign in the definition of the polar equation — and then press ENTER to toggle the equal sign between highlighted and unhighlighted. In the example in Figure 16-6, the calculator won't graph equations r_1, r_2, and r_4, but it will graph r_3.

Using ZOOM commands

After you've graphed your polar equations (as explained earlier in this chapter), you can use Zoom commands to change the viewing window of your graph. Pressing ENTER accesses the Zoom menu. On this menu you see all the Zoom commands that are available for graphing functions. What these commands do, and how you use them, is explained in detail in Chapter 10.

The Zoom commands that are most useful when graphing polar equations are:

- **ZoomFit:** This command finds a viewing window for a specified portion of the graph. How to use **ZoomFit** is explained in the "Graphing Polar Equations" section earlier in this chapter.

- **ZSquare:** Because there are more columns of pixels than rows on the calculator screen, circles will look like ellipses if the viewing window isn't properly set. **ZSquare** adjusts the settings in the Window editor so that circles look like circles. To use **ZSquare,** first graph your polar equations and then press ZOOM 5. The graph is then redrawn in a viewing window that makes circles look like circles. Figure 16-10 illustrates the effect that **ZSquare** has on the spiral $x = t\cos(t)$, $y = t\sin(t)$.

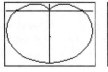

ZoomFit   ZSquare

Figure 16-10: A cardioid graphed using **ZoomFit** and then using **ZSquare**.

✔ **Zbox:** This command allows you to define a new viewing window for a portion of your graph by enclosing it in a box as illustrated in Figure 16-11. The box becomes the new viewing window. To construct the box:

1. **Press** ZOOM 1 .

2. **Define a corner of the box.**

   To do so, use the ▶ ◀ ▲ ▼ keys to move the cursor to the spot where you want one corner of the box to be located, and then press ENTER . The calculator marks that corner of the box with a small square.

3. **Construct the box.**

   To do so, press the ▶ ◀ ▲ ▼ keys. As you press these keys, the calculator draws the sides of the box.

   When you use **ZBox,** if you don't like the size of the box, you can use any of the ▶ ◀ ▲ ▼ keys to resize the box. If you don't like the location of the corner you anchored in Step 2, press CLEAR and start over with Step 1.

4. **When you're finished drawing the box, press** ENTER **and the graph will be redrawn in the viewing window specified by the box.**

   When you use **ZBox,** ENTER is pressed only two times. The first time you press it's to anchor a corner of the zoom box. The next time you press ENTER is when you're finished drawing the box and are ready to have the calculator redraw the graph.

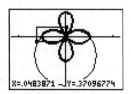

X=.0483871 _ Y=.37096774

Figure 16-11: Using **ZBox**.

✔ **Zoom In and Zoom Out:** After the graph is drawn, these commands allow you to zoom in on a portion of the graph or to zoom out from the graph. They work very much like a zoom lens on a camera. To use these commands, follow these steps:

1. **Press ZOOM 2 if you want to zoom in, or press ZOOM 3 if you want to zoom out.**

2. **Use the ▶ ◀ ▲ ▼ keys to move the cursor to the spot on the screen from which you want to zoom in or zoom out.**

   This spot becomes the center of the new viewing window.

3. **Press ENTER.**

   You can press ENTER again to zoom in closer or to zoom out one more time.

4. **Press CLEAR when you're finished zooming in or zooming out.**

If you used a Zoom command to redraw a graph and then want to undo what that command did to the graph, press ZOOM ▶ 1 to select **ZPrevious** from the Zoom Memory menu. **ZPrevious** redraws the graph as it appeared in the previous viewing window.

Saving a polar graph

After you've graphed your polar equations (as described earlier in this chapter), you can save the graph and its Window, Y= editor, Mode, and Format settings in a Graph Database. When you recall the Database at a later time, you get more than just a picture of the graph. The calculator also restores the Window, Y= editor, Mode, and Format settings to those stored in the database. So you can, for example, trace the recalled graph.

The procedures for saving and recalling a Graph Database are described in Chapter 9.

Tracing a polar graph

After you've graphed your polar equations (as described earlier in this chapter), press TRACE and then use ▶ to investigate these equations. Here's what you will see, and what you can do to change things:

❙ ✔ **The definition of the polar equation:** The polar equation you're tracing is displayed at the top of the screen, provided the calculator is in **ExprOn** format, as indicated in Figure 16-7. If you have graphed more than one polar equation and you would like to trace a different equation, press ▲. Each time you press this key, the cursor jumps to another polar equation. Eventually it jumps back to the original equation.

✔ **The independent variable θ:** The value of θ corresponding to the cursor location is displayed in the lower-left corner of the screen, provided the calculator is in **CoordOn** format, as indicated in Figure 16-7. When you press TRACE, the cursor is then placed at the beginning of the graph of the polar equation and θ displays the value you assigned to **θmin** in the Window editor. Each time you press ▶, the cursor moves to the next plotted point in the graph and the value of θ changes to the value of the independent variable corresponding to that plotted point.

If you press ◀, the cursor will move to the previously plotted point in the graph. And if you press ▲ to trace a different polar equation, the tracing of that equation will start at the same value of θ that was displayed on the screen before you pressed this key.

✔ **The values of _x_ and _y_:** At the bottom of the screen you see the values of the _x_- and _y_ coordinates of the cursor location, provided the calculator is in **RectGC** and **CoordOn** format (as shown previously in Figure 16-7). In the **PolarGC** format, the coordinates of this point display in polar form.

Press CLEAR to terminate tracing the graph. This also removes the name of the function and the coordinates of the cursor from the screen.

If your cursor disappears from the screen while you are tracing a polar graph, the sidebar "Panning in Polar mode" tells you how to rectify this situation.

If you're using TRACE and you want to start tracing your polar equation at a specific value of the independent variable θ, just key in that value and press ENTER. (The value you assign to θ must be between **θmin** and **θmax;** if it isn't, you will get an error message.) After you press ENTER, the trace cursor moves to the point on the graph corresponding to that value of θ. If that point is not on the part of the graph appearing on the screen, you can get the cursor and the graph in the same viewing window. The sidebar, "Panning in Polar mode," tells you how to do so.

If the name of the polar equation and the value of the independent variable θ are interfering with your view of the graph when you use TRACE, increase the height of the window by pressing WINDOW and decreasing the value of **Ymin** and increasing the value of **Ymax.**

Panning in Polar mode

When you're tracing a polar equation and the cursor hits an edge of the screen, if θ is less than θ**max** and you continue to press ▶, the coordinates of the cursor will be displayed at the bottom of the screen, but the calculator won't automatically adjust the viewing window. So you won't see the cursor on the graph. To rectify this, make a mental note of the value of θ and then press ENTER. The calculator redraws the graph centered at the location of the cursor at the time you pressed ENTER.

Unfortunately, after the graph is redrawn, the trace cursor is placed at the beginning of the first polar equation appearing in the Y= editor and may not appear on the screen. To get the trace cursor back on the part of the graph displayed in the new viewing window, key in the value of θ that you made a mental note of, and then press ENTER. Also, if you weren't tracing the first polar equation appearing in the Y= editor, use ▶ to place the cursor on the graph you *were* tracing. The trace cursor then appears in the middle of the screen and you can use ▶ to continue tracing the graph.

Displaying Equations in a Table

After you've entered the polar equations into the calculator (as described earlier in this chapter), you can have the calculator create a table of the values of the polar equations. To create a table, follow these steps:

1. **Highlight the equal sign of those polar equations in the Y= editor that you want to appear in the table.**

   Only polar equations defined in the Y= editor with a highlighted equal sign will appear in the table. To highlight or unhighlight an equal sign, press Y=, use the ▶◀▲▼ keys to place the cursor on the equal sign in the definition of the polar equation, and then press ENTER to toggle the equal sign between highlighted and unhighlighted.

2. **Press 2nd WINDOW to access the Table Setup editor (shown in Figure 16-12).**

Figure 16-12: The Table Setup editor.

3. Enter a number in TblStart and then press [ENTER].

TblStart is the first value of the independent variable θ to appear in the table. In Figure 16-12, **TblStart** is assigned the value 0.

To enter the number you've chosen for **TblStart**, place the cursor on the number appearing after the equal sign, press the number keys to enter your number, and then press [ENTER].

4. Enter a number in ΔTbl and then press [ENTER].

ΔTbl provides the increment for the independent variable θ. You usually want to set **ΔTbl** equal to a small number. In Figure 16-12, **ΔTbl** is assigned the value π/24.

To enter the number you've chosen for **ΔTbl**, place the cursor on the number appearing after the equal sign, press the number keys to enter your number, and then press [ENTER].

5. Set the mode for Indpnt and Depend.

To change the mode of either **Indpnt** or **Depend**, use the [▶][◀][▲][▼] keys to place the cursor on the desired mode, either **Auto** or **Ask**, and then press [ENTER].

To have the calculator automatically generate the table for you, put both **Indpnt** and **Depend** in **Auto** mode. The first table in Figure 16-13 was constructed in this fashion.

If you want to create a user-defined table in which you specify which values of the independent variable θ are to appear in the table and then have the calculator figure out the corresponding values of the polar equations, put **Indpnt** in **Ask** mode and **Depend** in Auto mode. How you construct the table is explained in Step 6. The second table in Figure 16-13 was constructed in this fashion.

For a user-defined table, you don't have to assign values to **TblStart** and **ΔTbl** in the Table Setup editor.

The other combinations of mode settings for **Indpnt** and **Depend** are not especially useful when you're dealing with polar equations.

Automatically   User-defined
generated

Figure 16-13: Tables for polar equations.

6. Press 2nd GRAPH to display the table.

When you display the table, the modes you set for **Indpnt** and **Depend** in Step 5 determine what you see on-screen — and what you can do with the table. Here's what you see and what you can do:

• **Here's the scoop for an automatically generated table:**

If **Indpnt** and **Depend** are both in **Auto** mode, then when you press 2nd GRAPH, the table is automatically generated. To display rows in the table beyond the last row on the screen, repeatedly press ▼ until they appear. And repeatedly press ▲ to display rows above the first row on the screen.

Each time the calculator redisplays a table with a different set of rows, it also automatically resets **TblStart** to the value of θ appearing in the first row of the newly displayed table. To return the table to its original state, press 2nd WINDOW to access the Table Setup editor, and then change the value the calculator assigned to **TblStart**.

If you're constructing a table for more than two polar equations, only the first two will appear on the screen. To see another the other equations, repeatedly press ▶ until they appear. This causes one or more of the first two equations to disappear. To see them again, repeatedly press ◀ until they appear.

• **Here's what you get for a user-generated table:**

If you put **Indpnt** in **Ask** mode and **Depend** in **Auto** mode so that you can generate your own table, then when you display the table, it should be empty. If it's not, clear the table as described in the next section.

In an empty table, key in the first value of the independent variable θ you want to appear in the table, and then press ENTER. The corresponding values of the polar equations automatically appear. Key in the next value of θ you want in the table and press ENTER, and so on. The values for θ that you place in the first column don't have to be in any specific order, nor do they have to be between θ**min** and θ**max**.

While displaying the table for polar equations, you can edit the definition of the equations without going back to the Y= editor. To do so, use the ▶ ◀ ▲ ▼ keys to place the cursor on the column heading for that polar equation you want to edit and then press ENTER. Edit the definition of the equation (editing expressions is explained in Chapter 1) and press ENTER when you're finished. The calculator automatically updates the table and the definition of the polar equation in the Y= editor.

The word ERROR appearing in a table indicates that the polar equation is either undefined or not a real number at the corresponding value of θ.

Clearing a user-defined table

After you've created a user-defined table, as described in the previous section, you can clear its contents by performing the following steps:

1. **Press 2nd WINDOW Table Setup editor and then set Indpend to Auto.**

2. **Press 2nd GRAPH to display an automatically generated table.**

3. **Press 2nd WINDOW and set Indpend back to Ask.**

4. **Press 2nd GRAPH to display an empty table.**

An automatically generated table cannot be cleared. To change the contents of such a table, change the values assigned to **TblStart** and Δ**Tbl** in the Table Setup editor.

Viewing the table and the graph on the same screen

After you've graphed your polar equations and created a table for the polar equations (as described earlier in this chapter), you can view the graph and the table on the same screen. To do so, follow these steps:

1. **Press MODE.**

2. **Put the calculator in G-T screen mode.**

   To do so, use the ▶◀▲▼ keys to place the cursor on **G-T** in the bottom right hand corner of the Mode menu and press ENTER to highlight it (as shown in Figure 16-14).

3. **Press GRAPH.**

   After you press GRAPH, the graph and the table appear on the same screen.

Figure 16-14: Setting the mode for viewing a graph and a table.

If you press any key used in graphing polar equations, such as [ZOOM] or [TRACE], the cursor becomes active on the graph side of the screen. To return the cursor to the table, press [2nd][GRAPH].

If you press [TRACE] and then use the [▶][◀][▲][▼] keys to trace the graph, the value of the independent variable θ corresponding to the cursor location on the graph will be highlighted in the table and the column for the polar equation you're tracing will appear next to it. If necessary, the calculator updates the table (as shown in Figure 16-15) so you can see that row in the table.

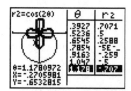

Figure 16-15: Using TRACE in Graph-Table mode.

To view the graph or the table in Full-Screen mode, follow these steps:

1. **Press** [MODE].

2. **Put the calculator in Full screen mode.**

   To do so, use the [▶][◀][▲][▼] keys to place the cursor on **Full** in the lower-left corner of the Mode menu and press [ENTER] to highlight it.

3. **Press** [GRAPH] **to see the graph, or press** [2nd][GRAPH] **to see the table.**

Evaluating Polar Equations

To evaluate a polar equation at a specified value of θ, follow these steps:

1. **Graph the equation in a viewing window that contains the specified value of θ.**

   Graphing polar equations and setting the viewing window are explained earlier in this chapter. To get a viewing window containing the specified value of θ, that value must be between θ**min** and θ**max**.

2. **Set the Format menu to ExprOn and CoordOn.**

   Setting the Format menu is explained earlier in this chapter.

3. **Press [2nd][TRACE] to access the Calculate menu.**

4. **Press [1] to select the value option.**

5. **Enter the specified value of θ.**

   To do so, use the keypad to enter the value of θ (as in the first graph in Figure 16-16). If you make a mistake when entering your number, press [CLEAR] and re-enter the number.

6. **Press [ENTER].**

   After you press [ENTER], the first highlighted polar equation in the Y= editor appears at the top of the screen, the cursor appears on the graph of that equation at the specified value of θ, and the coordinates of the cursor appear at the bottom of the screen. The second graph in Figure 16-16 illustrates this result.

7. **Repeatedly press [▲] to see the value of the other graphed polar equations at your specified value of θ.**

   Each time you press [▲] the name of the polar equation being evaluated appears at the top of the screen and the coordinates of the cursor location appear at the bottom of the screen, as in the third graph in Figure 16-16.

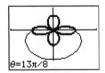 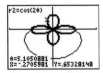

Enter θ   Press [ENTER]   Press [▲]

Figure 16-16: Steps in evaluating equations at a specified value of θ.

If you're planning to evaluate polar equations at several specified values of θ, consider constructing a user-defined table.

Finding Derivatives

To find the derivative (dy/dx or $dr/d\theta$) of a polar equation at a specified value of θ, follow these steps:

1. **Graph the polar equation in a viewing window that contains the specified value of θ.**

   Graphing polar equations and setting the viewing window are explained earlier in this chapter. To get a viewing window containing the specified value of θ, that value must be between θ**min** and θ**max.**

2. **Set the Format menu to ExprOn and CoordOn.**

   Setting the Format menu is explained earlier in this chapter.

3. **Press 2nd TRACE to access the Calculate menu.**

4. **Press 2 or 3 to respectively select the dy/dx or dr/dθ option.**

5. **If necessary, repeatedly press ▲ until the appropriate polar equation appears at the top of the screen.**

   This is illustrated in the first graph in Figure 16-17.

6. **Enter the specified value of θ.**

   To do so, use the keypad to enter the value of θ. As you use the keypad, θ= replaces the coordinates of the cursor location appearing at the bottom of the screen in the previous step. The number you key in is placed after θ = (as shown in the second graph in Figure 16-17). If you make a mistake when entering your number, press CLEAR and re-enter the number.

 If you're interested only in finding the slope of the curve in a general area of the graph — instead of at a specific value of θ — then just use the ◄ and ► to move the cursor to the desired location on the graph (instead of entering a value of θ).

7. **Press ENTER.**

   After pressing ENTER, the derivative is displayed at the bottom of the screen, as in the third graph in Figure 16-17.

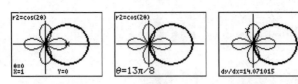

Select equation   Enter θ   Press ENTER

Figure 16-17: Steps in finding the *dy/dx* at a specified value of θ.

Part VI
Probability and Statistics

"Okay — let's play the statistical probabilities of this situation. There are 4 of us and 1 of him. Phillip will probably start screaming, Nora will probably faint, you'll probably yell at me for leaving the truck open, and there's a good probability I'll run like a weenie if he comes toward us."

In this part...

*T*his part gives you a look at calculating permutations and combinations, as well as generating random numbers. I also show you how to graph and analyze one- and two-variable statistical data sets. And if you want to do regression modeling (curve-fitting) — hey, who doesn't? — I show you how to do that, too.

Chapter 17

Probability

. . . .

In This Chapter

▶ Evaluating permutations using the Math Probability menu

▶ Evaluating combinations using the Math Probability menu

▶ Generating random numbers using (you guessed it) the Math
  Probability menu

. . . .

Permutations and Combinations

A *permutation,* denoted by **nPr,** answers the question: "From a set
of **n** different items, how many ways can you select *and* order
(arrange) **r** of these items?" A *combination*, denoted by **nCr,**
answers the question: "From a set of **n** different items, how many
ways can you select (independent or order) **r** of these items?" To
evaluate a permutation or combination, follow these steps:

1. **On the Home screen, enter n, the total number of items
   in the set.**

   If you're not already on the Home screen, press [2nd][MODE] to
   exit (quit) the current screen and enter the Home screen.

2. **Press [MATH][▶][▶][▶] to access the Math Probability menu.**

3. **Press [2] to evaluate a permutation or press [3] to evaluate
   a combination.**

4. **Enter r, the number of items selected from the set, and
   press [ENTER] to display the result.**

   Figure 17-1 illustrates this procedure. Notice that the last
   two lines in the figure tell you that it's impossible to select
   7 items from a set of only 5 items.

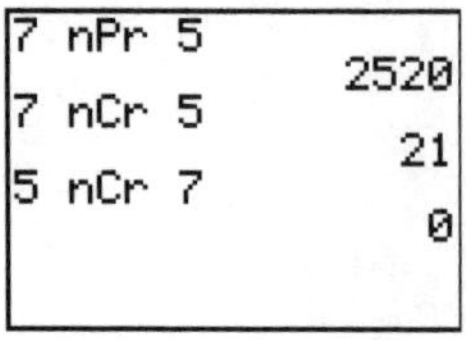

```
7 nPr 5
             2520
7 nCr 5
               21
5 nCr 7
                0
```

Figure 17-1: Evaluating permutations and combinations.

Generating Random Numbers

When generating random numbers, you usually want to generate numbers that are integers contained in a specified range, or decimal numbers that are strictly between 0 and 1.

Generating random integers

To generate random integers that fall between the integers **a** and **b**:

1. **Press** MATH▸▸▸5 **to select the** randInt **command from the Math Probability menu.**

2. **Enter the value of the lower limit** a, **press** , **, and enter the upper limit** b.

3. **Press** ENTER **to generate the first random integer. Repeatedly press** ENTER **to generate more random integers.**

   This is illustrated in the first picture in Figure 17-2.

Pressing any key except ENTER stops the calculator from generating random integers.

Generating random decimals

To generate random decimal numbers that are strictly between 0 and 1, press MATH▸▸▸1 to select the **rand** command from the Math Probability menu. Then repeatedly press ENTER to generate the random numbers. The second picture in Figure 17-2 illustrates this process.

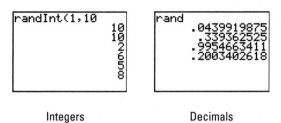

Integers   Decimals

Figure 17-2: Generating random numbers.

Chapter 18

Dealing with Statistical Data

● ●

In This Chapter

▶ Entering data into the calculator

▶ Deleting and editing data in a data list

▶ Saving and recalling a data list

▶ Sorting data lists

● ●

*T*he calculator has many features that provide information about
the data that has been entered in the calculator. It can graph it
as a scatter plot, histogram, or box plot. It can calculate the median
and quartiles. It can even find a regression model (curve fitting) for
your data. It can do this and much, much more. This chapter tells
you how to enter your data in the calculator; Chapter 19 shows you
how to use the calculator to analyze that data.

Entering Data

What you use to enter statistical data into the calculator is the Stat
List editor — a relatively large spreadsheet that can accommodate
up to 20 columns (data lists). And each data list (column) can handle
a maximum of 999 entries. Pictures of the Stat List editor appear in
Figure 18-1.

To use stat lists to enter your data into the calculator, follow
these steps:

1. Press STAT 5 ENTER **to execute the SetUpEditor command.**

The SetUpEditor command clears all data lists (columns)
from the Stat List editor and replaces them with the six
default lists L_1 through L_6. Any lists that are cleared from
the editor by this command are still in the memory of the
calculator; they just don't appear in the Stat List editor.

2. Press STAT 1 **to enter the Stat List editor.**

If no one has ever used the Stat List editor in your calculator, then the Stat List editor will look like the first picture in
Figure 18-1. If the Stat List editor has been used before,
then some of the default lists **L₁** through **L₆** may contain
data, as in the second picture in Figure 18-1.

L1	L2	L3	1
	------	------	

L1(1) =

L1	L2	L3	1
1890	-1.8	------	
1891	0		
1892	0		
1893	-.92		
1894	-4.63		
1895	-1.94		
1896	0		

L1(1)=1890

Empty lists   Lists with data

Figure 18-1: The Stat List editor.

**3. If necessary, clear lists L₁ through L₆ or create a user-
named list.**

The calculator requires that each data set (list) have a
name. You can use the default names **L₁** through **L₆**, or you
can create your own name for a data list. If you want to
create your own name, the section titled "Creating User-
Named Data Lists" tells you how.

If you want to use one of the default names **L₁** through **L₆**
for your data list but that list already contains data, then
you must first clear the contents of the list before you
enter new data into it. (The next section, "Deleting and
Editing Data," tells you how to clear a data list.)

4. Enter your data. Press ENTER **after each entry.**

Use the ▶◀▲▼ keys to place the cursor in the column
where you want to make an entry. Use the keypad to enter
your number and press ENTER when you're finished. A
column (list) can accommodate up to 999 entries.

To quickly enter data that follows a sequential pattern (such as 1,
2, 3, ..., 100 or 1, 4, 9, 25, ..., 1600, or 10, 20, ..., 200), see the section
titled "Using Formulas to Enter Data."

Deleting and Editing Data

Sooner or later you'll have to remove or modify the data that you've
placed in a data list. The following list shows you how to do so.

✔ **Deleting a data list from the memory of the calculator:**

To permanently remove a data list from the memory of the calculator, press 2nd + 2 to enter the Memory Management menu. Then press 4 to see the data lists that are stored in memory. Use ▼ to move the indicator to the list you want to delete, and then press DEL to delete that list. When you're finished deleting lists from memory, press 2nd MODE to exit (quit) the Memory Management menu and return to the Home screen.

Although the calculator does allow you to delete a default list (L_1 through L_6) from memory, in reality it deletes only the contents of the list and not its name.

✔ **Clearing the contents of a data list:**

When you clear a data list, the list's contents (and not its name) will be erased leaving an empty data list in the calculator's memory. To clear the contents of a data list in the Stat List editor, use the ▶◀▲▼ keys to place the cursor on the name of a list appearing in a column heading, and then press CLEAR ENTER.

✔ **Deleting a column (list) in the Stat list editor:**

To delete a column (list) from the Stat List editor, use the ▶◀▲▼ keys to place the cursor on the name of the list appearing in the column headings and then press DEL. The list will be removed from the Stat List editor but not from the memory of the calculator.

✔ **Deleting an entry in a data list:**

To delete an entry from a data list, use the ▶◀▲▼ keys to place the cursor on that entry, and then press DEL to delete the entry from the list.

✔ **Editing an entry in a data list:**

To edit an entry in a data list, use the ▶◀▲▼ keys to place the cursor on that entry, press ENTER, and then edit the entry or key in a new entry. If you key in the new entry, the old entry is automatically erased. Press ENTER when you're finished editing or replacing the old entry.

Creating User-Named Data Lists

To name a data list in the Stat List editor, follow these steps:

1. If necessary, press STAT 1 to enter the Stat List editor.

2. **Use the ▶◀▲▼ keys to place the cursor on the column heading where you want your user-named list to appear.**

   Your user-named list is created in a new column that appears to the left of the column highlighted by the cursor (as shown in the first picture in Figure 18-2).

3. **Press [2nd][DEL] to insert the new column.**

   The second picture in Figure 18-2 shows this procedure.

4. **Enter the name of your data list and press [ENTER].**

   The name you give your data list can consist of one to five characters that must be letters, numbers, or the Greek letter θ. The first character in the name must be a letter or θ.

   The Ⓐ after **Name** = indicates that the calculator is in Alpha mode. In this mode, when you press a key you enter the green letter above the key. To enter a number, exit the mode by pressing [ALPHA] again and then enter the number. To enter a letter after entering a number, you must press [ALPHA] to put the calculator back in Alpha mode (as in the third picture in Figure 18-2). Press [ENTER] when you're finished entering the name.

L1	▣	L3	2
1890	-1.8	------	
1891	0		
1892	0		
1893	-.92		
1894	-4.63		
1895	-1.94		
1896	0		

L2 ={-1.8,0,0, -....

Indicate column

L1	▬▬	L2	2
1890		-1.8	
1891		0	
1892		0	
1893		-.92	
1894		-4.63	
1895		-1.94	
1896		0	

Name=Ⓐ

Press [2nd] [DEL]

L1	▬▬	L2	2
1890		-1.8	
1891		0	
1892		0	
1893		-.92	
1894		-4.63	
1895		-1.94	
1896		0	

Name=YEAR

Enter name

Figure 18-2: Steps for creating a user-named data list.

If the name you give your data list is the name of a data list stored in memory, then after entering that name and pressing [ENTER], the data in the list stored in memory will be automatically entered in the Stat List editor.

After you have named your data list, you can press ▼ and start entering your data. Or, if appropriate, you can use a formula to generate the data. (See the next section, "Using Formulas to Enter Data.") And if the data you want to put in the newly named list is in another column of the Stat List editor — or in a list stored in memory under another name — you can paste that data into your newly named list. (See the section called "Saving and Recalling Data Lists," later in this chapter.)

Using Formulas to Enter Data

Figure 18-3 illustrates how you would place the sequence 10, 20, ..., 200 in list L_1. The formula used in this example is simply x. The initial and terminal values of x are naturally 10 and 200, respectively. And, as you may guess, x is incremented by 10.

To use a formula to define your data:

1. **If necessary, press** STAT 1 **to enter the Stat List editor.**

2. **Use the** ▶ ◀ ▲ ▼ **keys to place the cursor on the column heading where you want your data to appear and press** ENTER **.**

3. **Press** 2nd STAT ▶ 5 **to access the seq command.**

4. **Enter your formula as a function of a single variable, press** ⎡,⎤**, and then enter the name of the variable and press** ⎡,⎤**.**

   The first picture in Figure 18-3 shows this process; here the formula is x and the variable is x.

5. **Enter the initial value of your variable, press** ⎡,⎤**, enter the terminal value of your variable, press** ⎡,⎤**, enter the increment for the variable, and then press** ⎡)⎤**.**

   The second picture in Figure 18-3 shows this process.

6. **Press** ENTER **to enter your data in the calculator.**

   This procedure is shown in the third picture in Figure 18-3.

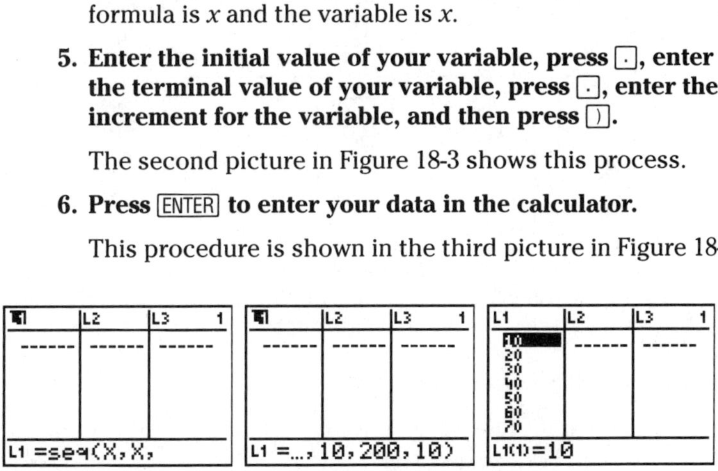

◼1	L2	L3	1		◼1	L2	L3	1		L1	L2	L3	1
------	------	------			------	------	------			10	------	------	
										20			
										30			
										40			
										50			
										60			
										70			
L1 =seq(X,X,					L1 =...,10,200,10)					L1(1)=10			

Figure 18-3: Steps for using a formula to define a data set.

Saving and Recalling Data Lists

Saving your data lists is the first step in the direction of calling them up again when you want to use or change them. The following list shows you how to do so:

✔ **Saving data lists:**

After you enter data into the Stat List editor, that data is automatically stored in the memory of the calculator under the list name that appears as the column heading for that list. You don't have to take any further steps to ensure that the calculator saves your data. However, if you clear the contents of a data list (as described in the section titled "Deleting and Editing Data"), the calculator retains the name of the data list in memory but deletes the contents of that list.

If you've entered your data in one of the default lists L_1 through L_6 and would like to save it as a named list, first create a user-named data list; you should get a result that resembles the first picture in Figure 18-4. Then press ENTER 2nd and press the number key for your data list (as in the second picture in Figure 18-4). Finally, press ENTER to insert the data in the default data list into the newly named data list. The third picture in Figure 18-4 shows this process.

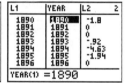

Create named list   Enter default list   Press ENTER

Figure 18-4: Steps for copying data from one list to another.

✔ **Recalling data lists:**

You can use the **SetUpEditor** command to set up the Stat List editor with the data lists you specify. To do this press STAT 5 to invoke the SetUpEditor command. Enter the names of the data lists, separated by commas. Then press ENTER STAT 1 to see the data lists (as shown in Figure 18-5).

If you're already in the Stat List editor, you can recall a saved data list by creating a data list that has the same name as the saved data list. (For pointers on how to do so, see the section of this chapter called "Creating User-Named Data Lists.")

```
SetUpEditor NUMS
,LONE,DEP
```

NUMS	LONE	DEP	1
1	21	32	
2	36	45	
3	44	50	
4	38	49	
5	52	60	
6	45	56	
7	29	34	

NUMS(1) = 1

Specify lists   View lists

Figure 18-5: Recalling saved data lists.

You can save a data list on your PC and then recall it at a later date. (Chapter 22 tells you how.) You can also transfer a data list from one calculator to another, as described in Chapter 23.

Sorting Data Lists

To sort data lists, follow these steps:

1. **Press** STAT.

2. **Press** 2 **to sort the list in ascending order, or press** 3 **to sort it in descending order.**

3. **Enter the list name.**

   To sort a default named list such as L_1, press 2nd 1 to enter its name. If you're sorting a user-named list, first press 2nd STAT ▶ ▲ ENTER to insert the letter **L** and then enter the name of the list. Inserting the letter **L** tells the calculator that what follows this letter is a data list, as shown in the first line in the first picture in Figure 18-6.

   If you want to sort rows according to the contents of the data list you entered in Step 3, then (after completing Step 3) enter the names of the other lists, separated by commas. For example, the third line in the first picture in Figure 18-6 is the command needed to rearrange the rows in Figure 18-5 so the data in the second column is in ascending order.

4. **Press**) ENTER **to sort the list, then press** STAT 1 **to view the sorted list.**

   The second picture in Figure 18-6 shows the rows in Figure 18-5 rearranged so the second column is in ascending order.

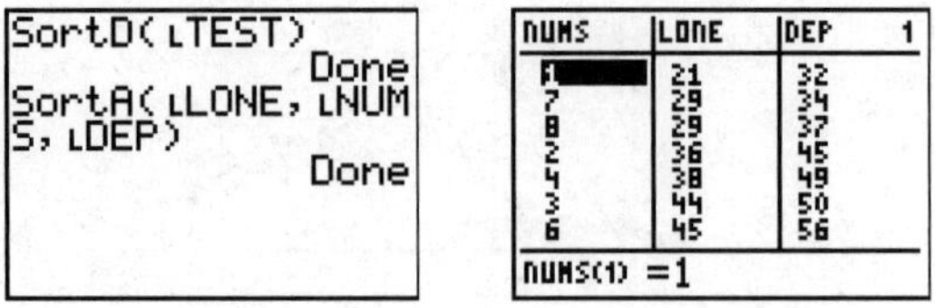

```
SortD(LTEST)
            Done
SortA(LLONE,LNUM
S,LDEP)
            Done
```

NUMS	LONE	DEP	1
1	21	32	
7	29	34	
8	29	37	
2	36	45	
4	38	49	
3	44	50	
6	45	56	

NUMS(1) =1

Sort command   Sorted data

Figure 18-6: Sorting data.

Chapter 19

Analyzing Statistical Data

. . . .

In This Chapter

▶ Plotting statistical data

▶ Creating histograms and box plots to describe one-variable data

▶ Creating scatter and line plots to describe two-variable data

▶ Tracing statistical data plots

▶ Finding the mean, median, standard deviation, and other neat stuff

▶ Finding a regression model for your data (curve fitting)

. . . .

*I*n descriptive statistical analysis, you usually want to plot your data and find the mean, median, standard deviation, and so on. You may also want to find a regression model for your data (a process also called *curve fitting*). This chapter tells you how to get the calculator to do these things for you.

Plotting One-Variable Data

The most common plots used to graph one-variable data are *histograms* and *box plots*. In a histogram, the data is grouped into classes of equal size; a bar in the histogram represents one class. The height of the bar represents the quantity of data contained in that class, as in the first picture in Figure 19-1.

A box plot (as in the second picture in Figure 19-1) consists of a box-with-whiskers. The box represents the data existing between the first and third quartiles. The box is divided into two parts, with the division line defined by the median of the data. The whiskers represent the locations of the minimum and maximum data points.

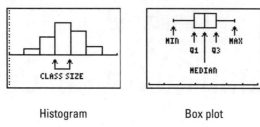

Histogram   Box plot

Figure 19-1: One-variable statistical plots.

Constructing a histogram

To construct a histogram of your data, follow these steps:

1. **Store your data in the calculator.**

   Entering data in the calculator is described in Chapter 18. Your data does not have to appear in the Stat List editor to plot it, but it does have to be in the memory of the calculator.

2. **Turn off any Stat Plots or functions in the Y= editor that you don't want to be graphed along with your histogram.**

   To do so, press Y= to access the Y= editor. The calculator graphs any highlighted Plots in the first line of this editor. To unhighlight a Plot so it won't be graphed, use the ▶◀▲▼ keys to place the cursor on the on the Plot and then press ENTER to toggle the Plot between highlighted and unhighlighted.

   The calculator graphs only those functions in the Y= editor defined by a highlighted equal sign. To unhighlight an equal sign, use the ▶◀▲▼ keys to place the cursor on the equal sign in the definition of the function, and then press ENTER to toggle the equal sign between highlighted and unhighlighted.

3. **Press 2nd Y= to access the Stat Plots menu and enter the number (1, 2, or 3) of the plot you want to define.**

   The first picture in Figure 19-2 shows this process, where **Plot1** is used to plot the data.

4. **Highlight On or Off.**

   If **On** is highlighted, the calculator plots your data. If you want your data to be plotted at a later time, highlight **Off**. To highlight an option, use the ▶◀▲▼ keys to place the cursor on the option, and then press ENTER.

Select plot number   Define plot   Press ZOOM 9

Figure 19-2: Steps for creating a histogram.

5. **Press ⊡, use ⊡ to place the cursor on the type of plot you want to create, and then press ENTER to highlight it.**

    Select ⅈⅉⅈ to construct a histogram.

6. **Press ⊡, enter the name of your data list (Xlist), and press ENTER.**

    If your data is stored in one of the default lists L_1 through L_6, press 2nd, key in the number of the list, and then press ENTER. For example, press 2nd 1 if your data is stored in L_1.

    If your data is stored in a user-named list, key in the name of the list and press ENTER when you're finished. Notice (as in Figure 19-2) that the calculator is already in Alpha mode, waiting for the first letter in the name of your list.

7. **Enter the frequency of your data.**

    If you entered your data without paying attention to duplicate data values, then the frequency is 1. On the other hand, if you did pay attention to duplicate data values, you most likely stored the frequency in another data list. If so, enter the name of that list the same way you entered the **Xlist** in Step 6.

8. **Press ZOOM 9 to plot your data using the ZoomStat command.**

    ZoomStat finds an appropriate viewing window for plotting your data, as in the third picture in Figure 19-2. The next step shows you how to redefine the class size in a histogram.

9. **Press WINDOW, set Xscl equal to the class size you desire, and then press GRAPH.**

    The class size (**Xscl**) must be ≥ (**Xmax** – **Xmin**)/47. If it isn't, you get the ERR: STAT error message.

10. **If necessary, adjust the settings in the Window editor.**

    The data plotted in Figure 19-2 consists of test scores. For such data, you naturally want the class size (**Xscl**) to be 10. But when the histogram is regraphed using this class size (as in the first picture in Figure 19-3), the viewing window doesn't accommodate the histogram. To correct this, adjust the settings in the Window editor (as described in Chapter 9). The second picture in Figure 19-3 shows the result of adjusting these settings.

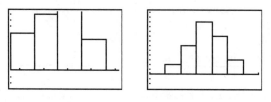

Adjust **Xscl**   Adjust window

Figure 19-3: Steps for defining the class size in a histogram.

Constructing a box plot

To construct a box plot for your data, follow Steps 1 through 8 for constructing a histogram. In Step 5, select the Box Plot symbol ⊞. If you adjust the viewing window as explained in Chapter 9, you can display a histogram and a box plot in the same viewing window (as shown in the first picture in Figure 19-4).

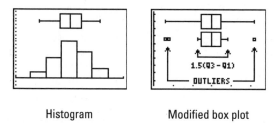

Histogram   Modified box plot

Figure 19-4: A box plot with a histogram and with a modified box plot.

If your data has *outliers*, consider constructing a modified box plot instead of a box plot. Figure 19-4 illustrates both a standard box plot and a modified box plot of the same data (in the second picture). In a modified box plot, the whiskers represent data in the range defined by 1.5(Q3 – Q1), and the outliers are plotted as points beyond the whiskers. The steps for constructing box plots and modified box plots are the same, except in Step 5 you select the modified box plot symbol ⊞.

Plotting Two-Variable Data

The most common plots used to graph two-variable data sets are the *scatter plot* and the xy-*line plot*. The scatter plot plots the points (*x*, *y*) where *x* is a value from one data list (**Xlist**) and *y* is the corresponding value from the other data list (**Ylist**). The *xy*-line plot is simply a scatter plot with consecutive points joined by a straight line.

To construct a scatter or *xy*-line plot, follow these steps:

1. **Follow Steps 1 through 6 in the previous subsection ("Constructing a histogram") with the following difference:**

   In Step 5, highlight ⌐∴ to construct a scatter plot or highlight ⌐⌁ to construct an *xy*-line plot.

2. **Enter the name of your Ylist and press** ENTER.

3. **Choose the type of mark used to plot points.**

   You have three choices: a small square, a small plus sign, or a dot. To select one, use ▶ to place the cursor on the mark and then press ENTER.

4. **Press** ZOOM 9 **to plot your data using the ZoomStat command.**

   ZoomStat finds an appropriate viewing window for plotting your data. This is illustrated in Figure 19-5.

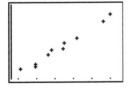   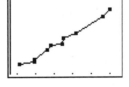

Scatter plot   *xy*-line plot

Figure 19-5: Two-variable statistical plots.

Tracing Statistical Data Plots

Before tracing a statistical data plot, press 2nd ZOOM and, if necessary, highlight the **CoordOn** in the second line of the Format menu and highlight **ExprOn** in the last line. This allows you to see the name of the data set being traced and the location of the cursor. To highlight an entry, use the ▶◀▲▼ keys to place the cursor on the entry and then press ENTER.

To trace a statistical data plot, press TRACE. In the upper left hand corner of the screen you see the Stat Plot number (P1, P2, or P3) and the name(s) of the data list(s) being traced. If you have more than one stat plot on the screen, repeatedly press ▲ until the plot you want to trace appears in the upper-left corner of the screen.

Use the ▶◀ keys to trace the plot. What you see depends on the type of plot.

✔ **Tracing a histogram:** As you trace a histogram, the cursor moves from the top center of one bar to the top center of the next bar. At the bottom of the screen you see the values of **min, max,** and **n.** This tells you that there are **n** data points x such that **min** $\leq x <$ **max**. This is illustrated in the first picture in Figure 19-6.

✔ **Tracing a box plot:** As you trace a box plot from left to right, the values that appear at the bottom of the screen are **minX** (the minimum data value), **Q1** (the value of the first quartile), **Med** (the value of the median), **Q3** (the value of the third quartile, and **maxX** (the maximum data value). This is illustrated in the second picture in Figure 19-6.

✔ **Tracing a modified box plot:** As you trace a modified box plot from left to right, the values that appear at the bottom of the screen are **minX** (the minimum data value), and then you see the values of the other outliers, if any, to the left of the interval defined by 1.5(Q3 – Q1). The next value you see at the bottom of the screen is the value of the left bound of the interval defined by 1.5(Q3 – Q1). Then, as with a box plot, you see the values of the first quartile, the median, and the third quartile. After that you see the value of the right bound of the interval defined by 1.5(Q3 – Q1), the outliers to the right of this, if any, and finally you see **maxX** (the maximum data value).

✔ **Tracing a scatter or *xy*-line plot:** As you trace a scatter plot or an *xy*-line plot, the coordinates of the cursor location appear at the bottom of the screen. As shown in Figure 19-6, the *x*-coordinate is a data value for the first data list named at the top of the screen; the *y*-coordinate is the corresponding data value from the second data list named at the top of the screen.

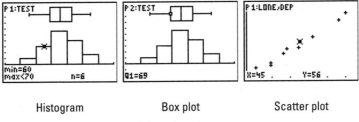

| Histogram | Box plot | Scatter plot |

Figure 19-6: Tracing statistical data plots.

Analyzing Statistical Data

The calculator can perform one- and two-variable statistical data analysis. For one-variable data analysis, the statistical data variable is denoted by **x**. For two-variable data analysis, the data variable for the first data list is denoted by **x** and the data variable for the second data list is denoted by **y**. Table 19-1 lists the variables calculated using one-variable data analysis (**1-Var**), as well as those calculated using two-variable analysis (**2-Var**).

Table 19-1   One- and Two-Variable Data Analysis

1-Var	*2-Var*	*Meaning*
$\bar{x}$	$\bar{x}$, $\bar{y}$	Mean of data values
Σx	Σx, Σy	Sum of data values
Σx^2	Σx^2, Σy^2	Sum of squares of data values
Sx	Sx, Sy	Sample standard deviation
σx	σx, σy	Population standard deviation
n	n	Total number of data points
minX	minX, minY	Minimum data value
maxX	maxX, maxY	Maximum data value
Q1		First quartile
Med		Median
Q3		Third quartile
	Σxy	Sum of x*y

One-variable data analysis

To analyze one-variable data, follow these steps:

1. **Store your data in the calculator.**

   Entering data in the calculator is described in the previous chapter. Your data does not have to appear in the Stat List editor to analyze it, but it does have to be in the memory of the calculator.

2. **Press** ⎡STAT⎤⎡▶⎤⎡1⎤ **to select the 1-Var Stats command from the Stat Calculate menu.**

3. **Enter the name of your data list (Xlist).**

   If your data is stored in one of the default lists L_1 through L_6, press ⎡2nd⎤, key in the number of the list, and then press ⎡ENTER⎤. For example, press ⎡2nd⎤⎡1⎤ if your data is stored in L_1.

   If your data is stored in a user-named list, first press ⎡2nd⎤⎡STAT⎤⎡▶⎤⎡▲⎤⎡ENTER⎤ to insert the letter **L** and then enter the name of the list. Inserting the letter **L** tells the calculator that what follows this letter is a data list. To enter the name of the list, use the ⎡ALPHA⎤ key to insert letters (as shown in the first picture of Figure 19-7).

4. **If necessary, enter the name of the frequency list.**

   If the frequency of your data is 1, you can skip this step and go to the next step. If, however, you stored the frequency in another data list, press ⎡,⎤ and then enter the name of that frequency list (just as you entered the **Xlist** in Step 3). The second picture in Figure 19-7 shows this process.

5. **Press** ⎡ENTER⎤ **to view the analysis of your data.**

   This is illustrated in the third picture in Figure 19-7. Use the ⎡▼⎤⎡▲⎤ keys to view the other values that don't appear on the screen.

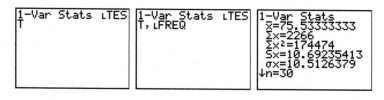

        Enter list   Enter frequency   Press ⎡ENTER⎤

Figure 19-7: Steps for one-variable data analysis.

Two-variable data analysis

To analyze two-variable data, follow these steps:

1. **Follow Steps 1 through 3 in the previous subsection ("One-variable data analysis") with the following difference:**

   In Step 2, press $\boxed{\text{STAT}}\boxed{\blacktriangleright}\boxed{2}$ to select the **2-Var Stats** command.

2. **Press $\boxed{,}$ and enter the name of the Ylist.**

3. **Follow Steps 4 and 5 in the previous subsection ("One-variable data analysis").**

Regression Models

Regression modeling is the process of finding a function that approximates the relationship between the two variables in two data lists. (An example appears in Figure 19-8, where a straight line approximates the relationship between the lists.) Table 19-2 shows the types of regression models the calculator can compute.

Table 19-2	Types of Regression Models	
TI-Command	*Model Type*	*Equation*
Med-Med	Median-median	$y = ax + b$
LinReg(ax+b)	Linear	$y = ax + b$
QuadReg	Quadratic	$y = ax^2 + bx + c$
CubicReg	Cubic	$y = ax^3 + bx^2 + cx + d$
QuartReg	Quartic	$y = ax^4 + bx^3 + cx^2 + dx + e$
LinReg(a+bx)	Linear	$y = a + bx$
LNReg	Logarithmic	$y = a + b*\ln(x)$
ExpReg	Exponential	$y = ab^x$
PwrReg	Power	$y = ax^b$
Logistic	Logistic	$y = c/(1 + a*e^{-bx})$
SinReg	Sinusoidal	$y = a*\sin(bx + c) + d$

To compute a regression model for your two-variable data, follow these steps:

1. **If necessary, turn on Diagnostics.**

   When the Diagnostics command is turned on, the calculator displays the correlation coefficient (r) and the coefficient of determination (r^2 or R^2) for appropriate regression models (as shown in the second picture in Figure 19-8). By default, this command is turned off. After you turn this command on, it stays on until you turn it off. Here's how to turn on Diagnostics:

   a. **Press [2nd][0][x^{-1}] to access the Catalog menu and to advance the Catalog to the entries beginning with the letter D.**

   b. **Repeatedly press [▾] to advance the Catalog indicator to the DiagnosticOn command.**

   c. **Press [ENTER] to paste this command on the Home screen, and press [ENTER] again to execute this command.**

      The first picture in Figure 19-8 shows this procedure.

2. **If necessary, put the calculator in Function (Func) mode.**

   If the regression model is a function that you want to graph, you must first put the calculator in Function mode. (Setting the mode is explained in Chapter 1.)

3. **If you haven't already done so, graph your two-variable data in a scatter or *xy*-line plot.**

   An earlier section of this chapter ("Plotting Two-Variable Data") explains how to do so.

4. **Select a regression model from the Stat Calculate menu.**

   To do so, press [STAT][▸] to access the Stat Calculate menu. Repeatedly press [▾] until the number or letter of the desired regression model is highlighted, and press [ENTER] to select that model.

5. **Enter the name for the Xlist data, press [,], and then enter the name of the Ylist data.**

   The appropriate format for entering list names is explained in Step 3 in the earlier subsection called "One-variable data analysis."

6. **If necessary, enter the name of the frequency list.**

   To determine whether to enter a frequency list, see Step 4 in the subsection earlier section called "One-variable data analysis."

7. **Press `,` and enter the name of the function (Y₁, ... , Y₉, or Y₀) in which the regression model is to be stored.**

   To enter a function name, press `VARS` `▶` `1` and then enter the number of the function (as in the first picture in Figure 19-8).

8. **Press `ENTER` to calculate and view the equation of the regression model.**

   This is illustrated in the second picture in Figure 19-8. The equation of the regression model is automatically stored in the Y= editor under the name you entered in Step 7.

9. **Press `GRAPH` to see the graph of your data and regression model.**

   This process is illustrated in the third picture in Figure 19-8.

Define model   Press `ENTER`   Press `GRAPH`

Figure 19-8: Steps for calculating and graphing a regression model.

Part VII
Dealing with Matrices

In this part...

In this part, I show you how to use matrices in arithmetic expressions and how to find the inverse, transpose, and determinant of a matrix. I also show you how to use matrices to solve a system of linear equations.

Chapter 20

Creating and Editing Matrices

. . . .

In This Chapter

▶ Defining a matrix

▶ Editing a matrix

▶ Displaying the contents of a matrix

▶ Augmenting two matrices

▶ Copying one matrix to another matrix

▶ Deleting a matrix from the memory of the calculator

. . . .

A *matrix* is a rectangular array of numbers arranged in rows and columns. With matrices, a system of equations can be easily manipulated or solved. The dimensions, r × c, of a matrix are defined by the number of rows and columns in the matrix. The calculator allows you to define up to ten matrices. Each matrix can have dimensions of up to 99 × 99 and can contain only real-number entries.

Defining a Matrix

To define a matrix, follow these steps:

1. **Press** 2nd x⁻¹ ▶ ▶ **to enter the Matrix editor menu. (On the TI-83, press** MATRX ▶ ▶ **.)**

   The first picture in Figure 20-1 shows the names of matrices used by the calculator. The dimensions appearing to the right of the first two matrices in this picture indicate that these two matrices have already been defined and are stored in the memory of the calculator.

2. **Key in the number (1 through 9, or 0) of the matrix you want to define or redefine.**

   If you select an already-defined matrix, the following steps redefine that matrix by overwriting any existing entries. The redefined matrix replaces the original matrix in the memory of the calculator.

   If all ten matrices in the Matrix editor are defined and you don't want to sacrifice any of them in order to define a new matrix, consider saving some of the already-defined matrices on your PC. Chapter 22 tells you how to do so.

3. **Enter the dimensions of the matrix. Press** ENTER **after each entry.**

   The dimensions, r × c, of a matrix indicate the number of rows and columns in the matrix. The number of rows is entered first.

4. **Enter the elements in the matrix. Press** ENTER **after each entry.**

   The calculator enters elements one row at a time. When you press ENTER after entering the last element in the first row, the calculator moves to the beginning of the second row and waits for you to make another entry.

   Before accepting an entry, the calculator displays the row and column of the entry at the bottom of the screen. It also displays the value assigned to that entry. To change that value, just key in the new value and press ENTER. The new value can also be entered as an arithmetic expression such as $9^2 + 7$. To leave that value as it is, just press ENTER, as in the second picture in Figure 20-1.

5. **When you're finished defining your matrix, press** 2nd MODE **to exit the Matrix editor and return to the Home screen.**

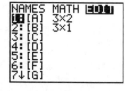   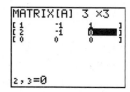

Select matrix   Define matrix

Figure 20-1: Steps for defining a matrix.

Editing a Matrix

To edit an already-defined matrix, follow these steps:

1. **Follow Steps 1 and 2 in the previous section.**

2. **Use the** ▶◀▲▼ **to place the cursor on the entry to be edited.**

3. **Key in the new number and press** ENTER.

4. **When you're finished editing your matrix, press** 2nd MODE **to exit (quit) the Matrix editor and return to the Home screen.**

Displaying Matrices

To display a matrix on the Home screen, first press 2nd MODE to access the Home screen. Then press 2nd x^{-1} and enter the number of the matrix to be displayed. (On the TI-83, press MATRX.) Finally, press ENTER to display the matrix, as in the first two pictures in Figure 20-2. If the Home screen isn't large enough to contain the whole matrix, use the ▶◀▲▼ keys to view the missing elements.

To view the contents of a matrix in the Matrix editor instead of on the Home screen, follow the first two steps in the "Defining a Matrix" section at the beginning of this chapter.

Augmenting Two Matrices

Augmenting two matrices allows you to append one matrix to another matrix. Both matrices must be defined and have the same number of rows. To augment two matrices, follow these steps:

1. **If necessary, press** 2nd MODE **to access the Home screen.**

2. **Press** 2nd x^{-1} ▶ 7 **to select the augment command from the Matrix Math menu. (On the TI-83, press** MATRX ▶ 7.**)**

3. **Enter the name of the first matrix and then press** ,.

   The first matrix is the matrix that appears on the left in the augmented matrix. To enter its name, press 2nd x^{-1} and key in the number of the matrix name. (On the TI-83, press MATRX.) This is illustrated in the third picture in Figure 20-2.

4. **Enter the name of the second matrix and then press**).

5. **Store the augmented matrix under a specified matrix name.**

To do so press STO►, enter the name of the matrix in which you plan to store the augmented matrix, and then press ENTER.

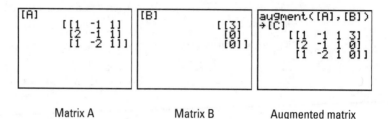

Matrix A   Matrix B   Augmented matrix

Figure 20-2: Augmenting two matrices.

Copying One Matrix to Another

There are several situations in which you may want to copy the contents of one matrix to another. One of these is when you want to define a new matrix by adding rows and columns to an existing matrix. To do so, copy the existing matrix to a new matrix and then redefine the new matrix to have the number of rows and columns you desire. Defining and redefining matrices is described in the first section of this chapter.

To copy one matrix to another matrix, follow these steps:

1. **If necessary, press 2nd MODE to access the Home screen.**

2. **Press 2nd x^{-1} and key in the number of the matrix you plan to copy. (On the TI-83, MATRX.)**

3. **Press STO►.**

4. **Press 2nd x^{-1} and key in the number of the matrix that will house the copy. (On the TI-83, MATRX.)**

If you copy the contents of a matrix to another matrix, the contents of that other matrix will be erased and replaced with the contents of the matrix you're copying. If all ten matrices in the Matrix editor are defined and you don't want to sacrifice any of them in order to make a copy of a matrix, consider saving some of the already defined matrices on your PC. Chapter 22 tells you how to do so.

5. **Press ENTER to save a copy of the matrix under the new name.**

The third picture in Figure 20-2 illustrates using STO▸ to save the contents of one matrix in another matrix.

Deleting a Matrix from Memory

To delete a matrix from the memory of the calculator, follow these steps:

1. **Press 2nd+25 to access the list of matrices that are in the memory of the calculator.**

2. **Repeatedly press ▾ to place the indicator next to the matrix you want to delete and press DEL.**

3. **When you're finished deleting matrices, press 2nd MODE to exit (quit) the Memory Manager and return to the Home screen.**

Chapter 21

Using Matrices

In This Chapter

▶ Using matrices in arithmetic expressions

▶ Finding a scalar multiple of a matrix

▶ Negating a matrix

▶ Using the identity matrix in an arithmetic expression

▶ Using matrices to solve a system of equations

▶ Converting a matrix to reduced row-echelon form

Matrix Arithmetic

When evaluating arithmetic expressions that involve matrices, you usually want to perform the following basic operations: scalar multiplication, negation (additive inverse), addition, subtraction, multiplication, and inversion (multiplicative inverse). You may also want to raise a matrix to an integral power or find its transpose. And you may want to use the identity matrix in an arithmetic expression.

Here's how you enter matrix operations in an arithmetic expression:

1. **Define the matrices in the Matrix editor.**

   This operation is explained in the first section of Chapter 20.

2. **Press** 2nd MODE **to access the Home screen.**

   All matrix operations are performed on the Home screen.

3. **If you want to clear the Home screen, repeatedly press** CLEAR**.**

4. **Enter the operations you want to perform and press** ENTER **when you're finished.**

   As with algebraic expressions, the Home screen is where you evaluate arithmetic expressions that involve matrices. To paste the name of a matrix into an expression, press 2nd x⁻¹ and key in the number of the matrix name. (On the

TI-83, press MATRX.) Here's how you enter the various operations into the arithmetic expression:

- **Entering the scalar multiple of a matrix:**

  To enter the scalar multiple of a matrix in an arithmetic expression, enter the value of the scalar and then enter the name of the matrix as shown in the first picture in Figure 21-1.

- **Negating a matrix:**

  To negate a matrix, press ($-$) and then enter the name of the matrix as shown in the second picture in Figure 21-1.

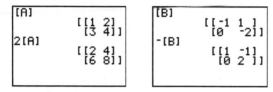

Scalar multiple   Negation

Figure 21-1: The scalar multiple and the negation of a matrix.

- **Entering the identity matrix:**

  You don't have to define an identity matrix in the Matrix editor in order to use it in an algebraic expression. To enter an identity matrix in an expression, press 2nd x⁻¹ ▸ 5 to select the **identity** command from the Matrix Math menu. (On the TI-83, press MATRX ▸ 5.) Then enter the size of the identity matrix. For example, enter **2** for the 2×2 identity matrix, as in the first picture in Figure 21-2.

- **Adding or subtracting matrices:**

  When adding or subtracting matrices, the matrices must have the same dimensions. If they don't, you will get the ERR: DIM MISMATCH error message.

  Entering the addition and subtraction of matrices is straightforward; just combine the matrices by pressing + or –, as appropriate. The second picture in Figure 21-2 illustrates this process.

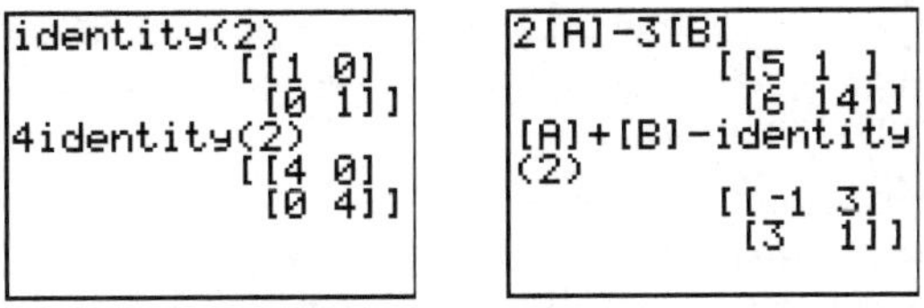

```
identity(2)
        [[1 0]
         [0 1]]
4identity(2)
        [[4 0]
         [0 4]]
```
```
2[A]-3[B]
        [[5 1 ]
         [6 14]]
[A]+[B]-identity
(2)
        [[-1 3]
         [3  1]]
```

Identity matrix   Adding and subtracting

Figure 21-2: The identity matrix and addition/subtraction of matrices.

- **Multiplying two matrices:**

    When finding the product A*B of two matrices, the number of columns in the first matrix A must equal the number or rows in the second matrix B. If this condition isn't satisfied, you will get the ERR: DIM MISMATCH error message.

    The multiplication of matrices is straightforward; just indicate the product by using juxtaposition or by pressing ⌧, as in the first picture in Figure 21-3.

- **Finding the inverse of a matrix:**

    When finding the inverse of a matrix, the matrix must be *square* (number of rows = number of columns) and *nonsingular* (nonzero determinant). If it is not square, you will get the ERR: INVALID DIM error message. If it is singular (determinant = 0), you will get the ERR: SINGULAR MAT error message. Evaluating the determinant of a matrix is explained in the next section.

    Entering the inverse of a matrix is straightforward; just enter the name of the matrix and then press [x⁻¹][WINDOW], as in the second picture of Figure 21-3.

- **Raising a matrix to a positive integral power:**

    When finding the power of a matrix, the matrix must be square. If it isn't, you will get the ERR: INVALID DIM error message.

    Only non-negative integers can be used for the power of a matrix. If the exponent isn't a nonnegative integer, you will get the ERR: DATA TYPE error message. If you raise a square matrix to the zero power, you will get the identity matrix.

TIP

To raise a square matrix to a negative power, raise the inverse of the matrix to the corresponding positive power.

Entering the positive power of a matrix is straightforward; just enter the name of the matrix, press $\land$, and enter the power, as in the third picture of Figure 21-3.

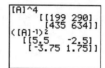

```
[A][C]
     [[5  -8   2]
      [9 -16  6]]
[A]*[C]
     [[5  -8   2]
      [9 -16  6]]
```
```
[A]⁻¹
     [[-2   1  ]
      [1.5  -.5]]
2[A]⁻¹-3[B]
     [[-1  -1]
      [3    5]]
```
```
[A]^4
     [[199 290]
      [435 634]]
<[A]⁻¹>²
     [[5.5    -2.5]
      [-3.75 1.75]]
```

Product of a matrix   Inverse of a matrix   Power of a matrix

Figure 21-3: The product and powers of matrices.

- **Transposing of a matrix:**

  To transpose a matrix in an arithmetic expression, enter the name of the matrix and then press [2nd][x⁻¹][▶][2] to select the Transpose command from the Matrix Math menu (on the TI-83, press [MATRX][▶][2].), as illustrated in the first picture in Figure 21-4.

Finding the Determinant

When finding the determinant of a matrix, the matrix must be square (number of rows = number of columns). If it isn't, you get the ERR: INVALID DIM error message.

To evaluate the determinant of a matrix, follow these steps:

1. **If necessary, press [2nd][MODE] to access the Home screen.**

2. **Press [2nd][x⁻¹][▶][1] to select the det command from the Matrix Math menu. (On the TI-83, press [MATRX][▶][1].)**

3. **Enter the name of the matrix and then press [ )].**

   To enter the name of the matrix, press [2nd][x⁻¹] and key in the number of the matrix name. (On the TI-83, press [MATRX].)

4. **Press [ENTER] to evaluate the determinant.**

   This procedure is illustrated in the second picture in Figure 21-4.

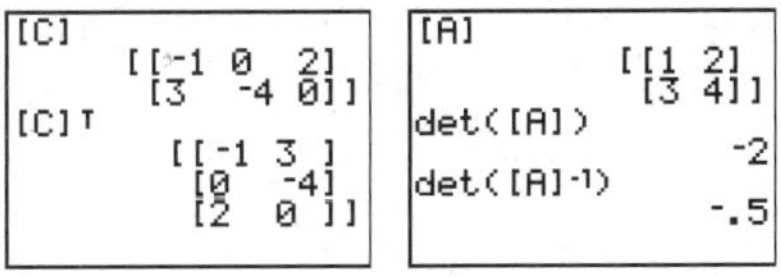

```
[C]
      [[-1  0   2]
       [3  -4   0]]
[C]ᵀ
      [[-1  3 ]
       [0  -4]
       [2   0 ]]
```

```
[A]
             [[1 2]
              [3 4]]
det([A])
                    -2
det([A]⁻¹)
                  -.5
```

Transpose   Determinant

Figure 21-4: The transpose and determinant of a matrix.

Solving a System of Equations

$$a_{11}x + a_{12}y + a_{13}z = b_1$$

$$a_{21}x + a_{22}y + a_{23}z = b_2$$

$$a_{31}x + a_{32}y + a_{33}z = b_3$$

Three matrices are associated with a system of linear equations: the coefficient matrix, the solution matrix, and the augmented matrix. For example, A, B, and C, are (respectively) the coefficient matrix, solution matrix, and augmented matrix for the system of equations just given.

$$A = \begin{bmatrix} a_{11} & a_{12} & a_{13} \\ a_{21} & a_{22} & a_{23} \\ a_{31} & a_{32} & a_{33} \end{bmatrix} \quad B = \begin{bmatrix} b_{11} \\ b_{21} \\ b_{31} \end{bmatrix} \quad C = \begin{bmatrix} a_{11} & a_{12} & a_{13} & b_1 \\ a_{21} & a_{22} & a_{23} & b_2 \\ a_{31} & a_{32} & a_{33} & b_3 \end{bmatrix}$$

Systems of linear equations can be solved by first putting the augmented matrix for the system in reduced row-echelon form. The mathematical definition of reduced row-echelon form isn't important here. It is simply an equivalent form of the original system of equations, which, when converted back to a system of equations, gives you the solutions (if any) to the original system of equations.

For example, when the reduced row-echelon matrix in the first picture in Figure 21-5 is converted to a system of equations, it gives the solutions $x = -3$, $y = 3$, and $z = 9$. The matrix in the second picture in Figure 21-5 converts to the system $x - z = 0$ and $y - z = -2$. This arrangement indicates that the system has an infinite number of solutions — namely, all solutions in which $x = z$ and $y = z - 2$, where z is any real number. The third picture in Figure 21-5 illustrates a system that has no solution — the last line of the matrix says that $0 = 1$, which is clearly impossible!

```
rref([C])
    [[1 0 0 -3]
     [0 1 0  3 ]
     [0 0 1  9 ]]
```
```
rref([D])
    [[1 0 -1  0 ]
     [0 1 -1 -2]
     [0 0  0  0 ]]
```
```
rref([E])
    [[1 1 0]
     [0 0 1]]
```

Unique solution   Infinite solutions   No solution

Figure 21-5: Converting a matrix to reduced row-echelon form.

To solve a system of equations, follow these steps:

1. **Define the augmented matrix in the Matrix editor.**

   The augmented matrix for the system of equations is explained at the beginning of this section. (Chapter 20 explains how to define a matrix in the Matrix editor.)

   You can define the coefficient and solution matrices for the system of equations and then augment these matrices to form the augmented matrix. (For more about augmenting matrices, see Chapter 20.)

2. **Press [2nd][MODE] to access the Home screen.**

3. **Press [2nd][x⁻¹][▶][ALPHA][APPS] to select the rref command. (On the TI-83, press [MATRX][▶][ALPHA][MATRX].)**

   You can also select the **rref** command by pressing [2nd][x⁻¹][▶] ([MATRX][▶] on the TI-83), repeatedly pressing [▾] until the cursor is next to the **rref** command, and then pressing [ENTER].

4. **Enter the name of the matrix and then press [ )].**

   To enter the name of the matrix, press [2nd][x⁻¹] and key in the number of the matrix name. (On the TI-83, press [MATRX].)

5. **Press [ENTER] to put the augmented matrix in reduced row-echelon form.**

6. **To find the solutions (if any) to the original system of equations, convert the reduced row-echelon matrix to a system of equations.**

   The beginning of this section describes converting a reduced row-echelon matrix to a system of equations.

Part VIII

Communicating with PCs and Other Calculators

The 5th Wave   By Rich Tennant

@RICHTENNANT

"You can sure do a lot with a TI-83 Plus, but I never thought dressing one up in G.I. Joe clothes and calling it your little desk commander would be one of them."

In this part...

*T*his part gets you ready to transfer files between your calculator and a PC, or between your calculator and another calculator. I also tell you how to download and install the free TI Connect software you can use to (among other things) transfer files to and from your PC.

Chapter 22

Communicating with a PC with TI Connect™

In This Chapter

▶ Downloading the TI Connect software

▶ Installing and running the TI Connect software

▶ Connecting your calculator to your computer

▶ Transferring files between your calculator and your computer

▶ Upgrading the calculator's operating system

*Y*ou need two things to enable your calculator to communicate with your computer: TI Connect (software) and a TI-Graph Link cable. TI Connect is free; the TI-Graph Link cable isn't. If it didn't come with your calculator, it can be purchased at the Texas Instruments online store at www.education.ti.com.

Downloading TI Connect

The TI Connect software is on the TI Resource CD that most likely came with your calculator. However, the version on this CD may not be current. The following steps tell you how to download the current version of TI Connect from the Texas Instruments Web site, as it existed at the time this book was published:

1. **Go to the Texas Instruments Web site at** www.education. ti.com.

2. **Click support at the top of the screen.**

3. **Click latest apps, software & operating system.**

4. **Click TI Connect.**

5. **Click downloads in the column to the left of the page.**

6. **Click the type of computer you use, Windows or Macintosh.**

7. **Click the FTP version of TI Connect.**

8. **Follow the directions given during the downloading process. Make a note of the directory in which you save the download file.**

   After you accept the License Agreement, you may be asked to log in. If you're not a member of the site, sign up — it's free.

Installing and Running TI Connect

After you've downloaded TI Connect, you install it by double-clicking the downloaded TI Connect file you saved in your computer. Then follow the directions given by the installation program you just launched.

After you start the TI Connect program, you see the many subprograms it contains. To see what these subprograms are used for, click the **HELP** button in the lower-right corner of the screen. In this chapter, I explain how to use TI Device Explorer to transfer files between your calculator and your PC.

Each of the subprograms housed in TI Connect have excellent Help menus that tell you exactly how to use the program.

Connecting Calculator and PC

You use the TI-Graph Link cable to connect your calculator to your computer. If you don't already have a TI-Graph Link cable, you can purchase one at the Texas Instruments online store at www. education.ti.com.

The TI-Graph Link cable comes in two flavors: serial or USB. Connect the serial or USB end of the cable to your computer and connect the other end to your calculator. The port for connecting the TI-Graph Link to the calculator is at the bottom of the calculator.

Transferring Files

After you've connected the calculator to your computer, the TI Device Explorer program housed in TI Connect can transfer files between the two devices. This allows you to archive calculator files on your computer.

To transfer files between your calculator and PC, start the TI Connect software and click the TI Device Explorer program. A directory appears, listing the files on your computer. Expanding this directory works the same as on your computer. When transferring files, you're usually interested in transferring the files housed in the following directories: Graph Database, List, Matrix, Picture, and Program. If any of these directories don't appear on-screen, that means no files are housed in that directory.

To copy or move files from your calculator to your PC, highlight the files you want to transfer, click **Action,** and select either **Copy to PC** or **Move to PC.** When the **Browse for Folder** window appears, select the location to which your files will be transferred and click **OK.**

To copy files to the calculator from a PC running Windows, you don't need to be in the TI Device Explorer program. Just open Windows Explorer, highlight the files you want to copy, right-click the highlighted files, select **Send To,** and click **Connected TI Device.** When asked if you want the files sent to **RAM** or **Archive,** select **RAM.** Files stored in the Archive memory of the calculator cannot be executed or edited. Directions for transferring files from a Macintosh to the calculator can be found in the TI Device Explorer Help menu.

The Help menu in TI Device Explorer is packed with useful information. In it you will find directions for editing and deleting calculator files and directions for backing up all files on your calculator.

Upgrading the OS

The TI-83 Plus and the TI-84 use Flash technology; the TI-83 does not. Flash technology allows you to upgrade the operating system of the calculator and to add application programs to your calculator.

Texas Instruments periodically upgrades the operating systems of the TI-83 Plus and the TI-83 Plus Silver Edition. To perform this upgrade, start the TI Connect software and click the TI Device Explorer program. Click the Help menu at the top of the screen and select **TI Device Explorer Help.** Click **Upgrading the device operating system** and follow the on-screen directions.

Chapter 23

Communicating Between Calculators

. . . .

In This Chapter

▶ Linking calculators so files can be transferred between them

▶ Determining what files can be transferred

▶ Selecting files to be transferred

▶ Transferring files between calculators

. . . .

*Y*ou can transfer data lists, programs, matrices, and other such files from one calculator to another if you link the calculators via the unit-to-unit link cable that came with your calculator. This chapter describes how to make such transfers.

Linking Calculators

Calculators are linked using the unit-to-unit link cable that came with the calculator. If you're no longer in possession of the cable, you can purchase one at the Texas Instruments online store at www.education.ti.com.

The unit-to-unit link cable connects to the calculator at the port at the bottom of the calculator. Push the cable into the port until you hear it click in place.

If you get the Error in Xmit error message when transferring files from one calculator to another, the most likely cause is a unit-to-unit link cable that wasn't fully inserted into the calculator port.

Transferring Files

If your calculators are the same, you can transfer anything between them. And you can transfer anything between a TI-83 Plus and a TI-83 Plus Silver Edition. You can also transfer anything from a TI-83 to the TI-83 Plus and the TI-83 Plus Silver Edition. But because the TI-83 is not endowed with Flash technology, you cannot transfer applications from a TI-83 Plus or TI-83 Plus Silver Edition to a TI-83.

After connecting two calculators, you can transfer files from one calculator (the sending calculator) to another (the receiving calculator). To select and send files, follow these steps:

1. **Press 2nd X,T,Θ,n on the sending calculator to access the Link Send menu.**

   The Link Send menu appears in the first picture in Figure 23-1.

2. **Use ▼ ▲ to select the type of file you want to send, and then press ENTER.**

   The first picture in Figure 23-1 shows the types of files you can send. The down arrow visible after number 7 in this list of menu items indicates that there are more menu items than can be displayed on-screen. Repeatedly press ▼ to view these other menu items.

   If you want to send all files on the calculator to another calculator, select **All+** and proceed to Step 4. (This option cannot be used when transferring files from a TI-83 Plus to a TI-83.)

3. **Use ▼ ▲ to move the cursor to a file you want to send and press ENTER to select that file. Repeat this process until you have selected all the files in this list that you want to send to another calculator.**

   The calculator places a small square next to the files you select, as in the second picture of Figure 23-1. In this picture, lists L_1, L_2, and L_3 are selected in the List Select menu.

4. **On the receiving calculator press 2nd X,T,Θ,n ▶ ENTER.**

   You see a screen that says **waiting,** and in its upper-right corner, a moving broken line indicates that the receiving calculator is waiting to receive files.

   Always put the receiving calculator in Receiving mode before you transfer files from the sending calculator.

```
SEND  RECEIVE
1:All+...
2:All-...
3:Prgm...
4:List...
5:Lists to TI82...
6:GDB...
7↓Pic...
```
```
SELECT  TRANSMIT
  L1        LIST
  L2        LIST
  L3        LIST
▶ L4        LIST
  L5        LIST
  L6        LIST
  FREQ      LIST
```

Link Send menu   List Select menu

Figure 23-1: Selecting files for transmission between calculators.

5. **On the sending calculator press ▶ ENTER to send the files to the receiving calculator.**

   As files are transferred, you may receive the **DuplicateName** menu, as illustrated in the first picture of Figure 23-2. This indicates that the receiving calculator already contains a file with the same name. Because the default names for stat lists are stored in the calculator, you always get this message when transferring a list, even if the list on the receiving calculator is empty.

   When you get the **DuplicateName** menu, select the appropriate course of action:

   • If you select **Overwrite,** any data in the existing file is overwritten by the data in the file being transferred.

   • If you select **Rename,** a new file is created and stored under the name you specify, as in the second picture in Figure 23-2.

   The 🅰 after **Name** = indicates that the calculator is in Alpha mode: When you press a key, what you enter is the green letter above the key. To enter a number, press ALPHA to take the calculator out of Alpha mode and then enter the number. To enter a letter after entering a number, you must press ALPHA to put the calculator back in Alpha mode. Press ENTER when you're finished entering the name.

   When renaming a file that is being transferred to the receiving calculator, the calculator has a strange and confusing way of warning you that a file having the same name already exists on the receiving calculator. When you press ENTER after entering the name, the calculator erases the name and makes you start over with entering a name. No warning message tells you that a file having the same name already exists on the calculator. If this happens to you, simply enter a different name.

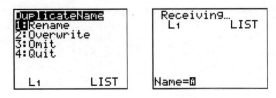

DuplicateName menu   Renaming a file

Figure 23-2: Dealing with duplicate file names.

Figure 23-3 illustrates a completed transfer of files between two calculators. The first picture in the figure shows the files that were sent; the second picture shows the files that were received. During the transfer of the files, L_2 was renamed **DATA.**

Sending calculator   Receiving calculator

Figure 23-3: Transferring files between two calculators.

If you want to terminate the transfer of files while it is in progress, press ON on either calculator. Then press ENTER when you're confronted with the Error in Xmit error message. If you put one calculator in Receiving mode and then decide not to transfer any files to that calculator, press ON to take it out of Receiving mode.

Transferring Files to Several Calculators

After transferring files between two calculators (as described in the previous section), you can then use the sending and/or receiving calculator to transfer the same files to a third calculator, usually without having to reselect the files. (If the initial transfer consisted of files selected from the **A-** submenu of the Link Send menu, then you will have to reselect the files.)

To transfer files to a third calculator, follow these steps:

1. **After making the initial transfer of files between the sending and receiving calculators, wait until Step 4 below before pressing any keys on the calculator that will be used to transfer the files to a third calculator.**

   After the initial transfer of files is complete, the screens on the sending and receiving calculators look similar to those in Figure 23-4, only with different files. If you press any key on the calculator, other than those specified in Step 3, the files you're planning on sending to a third calculator will no longer be selected and you will have to reselect them.

2. **Link the third calculator to either the sending or receiving calculator.**

3. **On the third calculator press [2nd][X,T,Θ,n][▸][ENTER].**

4. **On the other calculator press [2nd][X,T,Θ,n] and select the same menu item that was used in the initial transfer of files. Press [▸][ENTER] to complete the transfer of the files to the third calculator.**

   The files from the previous transfer are still selected, provided that in the interim you made no new selection from the Link Send menu (as shown in Figure 23-4). The selected files in this figure are the files that were sent to the receiving calculator in the initial transfer of files (refer to Figure 23-3).

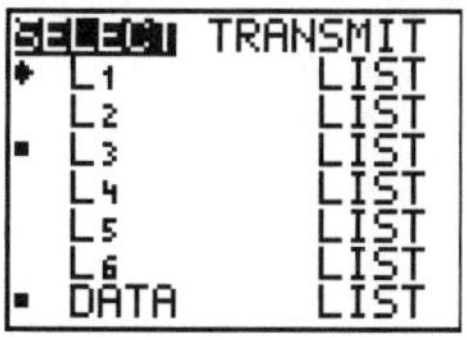

Figure 23-4: Transferring files to a third calculator.

Part IX
The Part of Tens

The 5th Wave   By Rich Tennant

I'm mathematically dyslexic. But it's not that unusual - 100 out of every 15 people are.

In this part...

*T*his part wraps up some handy items in packages of ten. I tell you how to download and install application programs that enhance the capabilities of your calculator, and briefly describe ten of my favorites. I also list the most common errors that crop up while using the calculator, and explain the most common error messages the calculator may give you.

Chapter 24

Ten Great Applications

In This Chapter

▶ Reviewing ten applications for the calculator

▶ Directions for finding and downloading applications

▶ Directions for installing applications on your calculator

*T*he TI-83 Plus and the TI-84 use Flash technology, the TI-83 does
not. Flash technology allows you to (among other things) install
application programs on your calculator.

If your calculator uses Flash technology, you're in for a treat
because the Texas Instruments Web site contains over 40 applications that you can download and install on your calculator. Most
of these application programs are free; those that aren't are very
inexpensive.

Texas Instruments may have already installed some of these programs on you calculator, even those that aren't free. To see what
application programs are already on your calculator, press APPS.

If your calculator doesn't have an APPS key, then you cannot add
application programs to the calculator because it isn't endowed
with Flash technology. If you want the apps, I recommend that you
upgrade to a TI-84.

In the following section, I briefly describe some application programs you can download from the Texas Instruments Web site. At
the end of this chapter I tell you how to find, download, and install
those programs.

Ten Great Applications

Here are ten great applications you can download from the Texas Instruments Web site:

- ✔ **Advanced Finance:** This program turns your calculator into a financial calculator.

- ✔ **Cabri Jr.:** This is an interactive, dynamic, geometry program that allows you to export geometric figures between the calculator and the Cabri Geometry II Plus software for Windows.

- ✔ **Catalog Help:** This handy program gives you the syntax for entering the argument required by a calculator function.

- ✔ **CellSheet:** This program turns your calculator into a spreadsheet. If you also download the free TI CellSheet Converter software, then you can transfer your spreadsheet files between your calculator and Microsoft Excel or AppleWorks.

- ✔ **GeoMaster:** This is a geometry program that will draw circles, triangles, and polygons. It will also make measurements and perform geometric transformations.

- ✔ **Inequality Graphing:** As the name implies, this program will graph inequalities.

- ✔ **NoteFolio:** This program turns your calculator into a word processor. Because using the calculator keys is more difficult than typing on a keyboard, I suggest purchasing a TI Keyboard so you can key in words the way you do on a computer. And don't forget to download the NoteFolio Creator software so you can transfer files between the calculator and Microsoft Word.

- ✔ **Organizer:** This program is a personal organizer in which you can schedule events, create to-do lists, and save phone numbers and e-mail addresses.

- ✔ **Probability Simulation:** This program simulates rolling dice, tossing coins, spinning a spinner, or drawing a card from a deck of cards.

- ✔ **StudyCards:** This program creates electronic flash cards. Be sure also to download the free TI StudyCard Creator software that allows you to create the flash cards on your PC.

Downloading an Application

The following steps tell you how to download application programs from the Texas Instruments Web site, as it existed at the time this book was published. To download and install applications, follow these steps:

1. **Go to the Texas Instruments Web site that is located at** www.education.ti.com.

2. **In the "Know what you're looking for?" drop-down menu, select "Apps & OS versions."**

3. **Click the type of calculator you have.**

4. **Click the application you want to download.**

5. **Click Download Instructions and read them.**

   The Download Instructions are the same for each application. So you need read them only once.

6. **Click the Guidebook to download it.**

   Save the application Guidebook (manual) on your computer, or print it.

7. **Click Download under the picture of the calculator and follow the directions you're given. Make a note of the directory in which you save the file you download.**

   After you accept the License Agreement, you're asked to Log In. If you aren't a member of the site, sign up — it's free.

Installing an Application

To install applications on your calculator you need the TI Connect software and a TI-Graph Link cable. Downloading and installing the software and connecting your calculator to your PC using the TI-Graph Link cable are explained in Chapter 22. Directions for copying the application file to your calculator are also found in Chapter 22.

Chapter 25

Eight Common Errors

• •

In This Chapter

▶ Incorrectly negating numbers

▶ Improperly indicating the order of operations

▶ Forgetting you're in Radian mode

▶ Graphing Stat Plots when functions are active and vice versa

• •

*E*ven the best calculating machine is only as good as its input. This chapter identifies eight common errors made when using the calculator.

Using ⊟ Instead of ⊡ to Indicate That a Number Is Negative

If you press ⊟ instead of ⊡ at the beginning of an entry, the calculator assumes you want to subtract what comes after the minus sign from the previous answer. If you use ⊟ instead of ⊡ in the interior of an expression to denote a negative number, the calculator responds with the ERR: SYNTAX error message.

Indicating the Order of Operations Incorrectly by Using Parentheses

When evaluating expressions, the order of operations is crucial. To the calculator, for example, -3^2 equals -9. This may come as quite a surprise to someone expecting the more standard evaluation, where $(-3)^2$ equals 9. The calculator first performs the operation in parentheses, then it squares the number, and after that, it performs the unary operation of negating a number. Unless you're careful, this won't provide the answer you're looking for. To guard against this error, you may want to review the detailed list of the order in which the calculator performs operations (given in Chapter 2).

Also, when graphing rational functions, users who are new to the calculator often make the basic mistake of omitting the parentheses that must be used to set the numerator apart from the denominator.

Improperly Entering the Argument for Menu Functions

If an argument is improperly entered, a menu function won't work. A prime example is the **fMin** function housed in the Math MATH menu. Do you remember what to place after this function so you can use it? If you don't, you get the ERR: ARGUMENT error message.

Texas Instruments offers a really invaluable (and free!) application called Catalog Help that you can install on your calculator. It reminds you of the appropriate argument that each function requires. I highly recommend installing it on your calculator.

Entering an Angle in Degrees While in Radian Mode

Actually, you *can* do so legitimately, but you have to let the calculator know that you're overriding the Angle mode by placing a degree symbol after your entry. Chapter 3 tells you how to do so.

Graphing Trigonometric Functions While in Degree Mode

This, too, is a mistake unless you do it just right: In the Window editor you have to set the limits for the x-axis as $-360 \le x \le 360$. Pressing ZOOM 7 or ZOOM 0 to have the calculator graph the function using the **ZTrig** or **ZoomFit** command produces similar results. But this works when you're graphing pure trig functions such as $\sin x$. If you're graphing something like $\sin x + x$, life is a lot easier if you graph it in Radian mode.

Graphing Functions When Stat Plots Are Active

If you get the ERR: INVALID DIM error message when you graph a function, this is most likely caused by a Stat Plot that the calculator is trying to graph along with your function. Chapter 9 tells you how to inactivate that Stat Plot.

Graphing Stat Plots When Functions or Other Stat Plots Are Active

If you get the ERR: INVALID DIM error message when you graph a Stat Plot, this is most likely caused by a function or another Stat Plot that the calculator is trying to graph along with your function. Chapter 19 tells you how to inactivate those functions or Stat Plots.

Setting the Window Inappropriately for Graphing

If you get the ERR: WINDOW RANGE error message when graphing functions, this is most likely caused by setting **Xmin** $\geq$ **Xmax** or by setting **Ymin** $\geq$ **Ymax** in the Window editor. Setting the Window editor is explained in Chapter 9.

Chapter 26

Eleven Common Error Messages

In This Chapter

▶ Listing the eleven most common error messages

▶ Getting familiar with the eleven most common error messages

▶ Avoiding the eleven most common errors (but you knew that)

*H*ere's a list of eleven (ten with an extra one thrown in for good measure) common error messages the calculator may give you:

ARGUMENT

You usually get this message when you are using a function housed in one of the menus on the calculator. This message indicates that you have not properly defined the argument needed to use the function.

Texas Instruments has a really invaluable application called Catalog Help that you can install on your calculator and use to remind you of the appropriate argument required by these functions. It's free, so I highly recommend that you install it on your calculator (as explained in Chapter 27).

BAD GUESS

This message indicates that the guess you've given to the calculator isn't within the range of numbers that you specified. This is one of those times when the calculator asks you to guess the solution. One example is when you're finding the maximum value or the zero of a function within a specified range (see Chapter 11). Another is when you're finding the solution to an equation where that solution in contained in a specified range (see Chapter 5).

One other time that you can get this message is when the function is undefined at (or near) the value of your guess.

DATA TYPE

This type of error occurs if, for example, you enter a negative number when the calculator requires a positive number.

DIM MISMATCH

You usually get this message when you attempt to add, subtract, or multiply matrices that don't have compatible dimensions.

DOMAIN

You usually get this message when you're using a function housed on a menu of the calculator. If that function is, for example, expecting you to enter a number in a specified range, you get this error message if that number isn't in the specified range.

INVALID

This is the catchall error message. Basically it means that you did something wrong when defining something (for example, you used function Y_3 in the definition of function Y_2, but forgot to define function Y_3).

INVALID DIM

You get this invalid-dimension message if (for example) you attempt to raise a nonsquare matrix to a power or enter a decimal for an argument of a function when it's expecting an integer.

NO SIGN CHNG

When you're using the Equation Solver (detailed in Chapter 5) you get this message when the equation has no real solutions in your specified range. When using the Finance application (discussed in Part III) you get this message when you don't use the correct sign for cash flow (for a refresher, see Chapter 6).

SINGULAR MAT

You get this message when you try to find the inverse of a matrix whose determinant is zero.

SYNTAX

This is another catchall error message. It usually means you have a typo somewhere.

WINDOW RANGE

This, of course, means that the Window in improperly set. This problem is usually (but not always) caused by improperly setting Xmin ≥ Xmax or Ymin ≥ Ymax in the Window editor. For a look at the proper way to set the Window for functions, sequences, parametric equations, and polar equations, check out the explanations in (respectively) Chapters 9, 13, 15, and 16.

Index

• Symbols and Numerics •

: (colon), 27
ΔTbl value, 133, 150
π key, 22
ΣInt command, 60
ΣPrn command, 60
θ (angular direction), 157
θmax variable
    (Polar mode), 162
θmin variable
    (Polar mode), 162
θstep variable
    (Polar mode), 162–163
2nd (secondary function) key, 10, 11

• A •

a + b*i* mode (Mode menu), 18, 41–42
aborting process, 13, 163
abs function (Math menu)
  CPX submenu, 43
  NUM submenu, 35
absolute value, finding, 35
accessing
  Angle menu, 157
  CPX submenu (Math menu), 42
  Format menu, 71
  Func (Function) mode
      (Mode menu), 69
  letters from keyboard, 11
  Math menu, 14, 31
  MATH submenu (Math menu), 33
  Matrix editor menu, 205
  menu, 14–15

Parametric mode
    (Mode menu), 140
Pol (Polar) setting
    (Mode menu), 160
Seq (Sequential) mode
    (Mode menu), 118
Stat Plots menu, 192
Test menu, 14
Y= editor, 140
Zoom menu, 127
accuracy of graph, 76–77
adding matrices, 212–213
Advanced Finance application, 232
Alpha key, 11
Alpha mode and Catalog, 19
analyzing statistical data, 197–199
angle function (Math menu,
    CPX submenu), 43
Angle menu
  accessing, 157
  converting between degrees and
      DMS, 38–39
  converting degrees to radians,
      37–38
  converting radians to degrees, 38
  entering angles in DMS
      measure, 39
  overriding mode of angle, 39–40
  overview of, 36–37
angles
  entering in degrees in Radian
      mode, 236
  entering in DMS measure, 39
  entering in radian measure, 159
  overriding mode of, 39–40
angular direction (θ), 157
Animate style, 74, 145

answer, starting expression with last, 24–25
applications
  downloading, and battery low warning, 9–10
  downloading from Texas Instruments Web site, 233
  Flash technology and, 231
  installing, 233
  on Texas Instruments Web site, 232
Apps key, 231
arguments
  defining, 239
  entering, 23–24, 236
Arrow keys, 10, 12
augment command, 207
augmenting two matrices, 207–208
automatically generated table
  changing contents of, 91, 152
  functions and, 90–91
  parametric equations and, 151
  polar equations and, 171
  sequences and, 134
AxesOn and AxesOff options (Format menu), 71

• B •

backup battery, 10
bal command, 61
balance of loan, finding, 61
battery low warning, 9–10
bound variable (Equation Solver), 47–48
box plots
  constructing, 194
  description of, 191–192
  tracing, 196, 197
busy indicator, 13

• C •

cable
  TI-Graph Link, 219, 220
  unit-to-unit link, 223
Cabri Jr. application, 232
Calculate (CALC) menu
  definite integral, evaluating, 102–103
  dy/dx, dy/dt, or dx/dt options, 155–156
  maximum or minimum points on graph, finding, 98–99
  point of intersection, finding, 99–100
  slope of curve, finding, 100–101
  value of function, finding, 95–97
  zeros of function, finding, 97–98
cash flow, distinguishing negative from positive, 56
cash-flow frequency list, 56
Catalog
  DiagnosticOn command, 200
  using, 18–19
Catalog Help application, 232, 236, 239
CellSheet application, 232
changing batteries, 10
circle, drawing on graph, 107–108
Clear key, 12, 13
clearing
  data list, 185
  user-generated table, 91–92
ClrDraw command, 102, 113
colon (:), 27
column, deleting in Stat List editor, 185
combination (nCr), 179–180
combining expressions, 27–28

commands. *See also* Zoom
  commands
augment, 207
bal, 61
ClrDraw, 102, 113
DiagnosticOn, 200
functions and, 76
Quit, 13
recalling last, 14
reusing, 54
rref, 216
SetUpEditor, 183, 188
ΣInt, 60
ΣPrn, 60
Transpose, 214
value, 96–97, 138
ZDecimal, 84
ZInteger, 85
ZPrevious, 129, 148
ZSquare, 84–85, 146, 165–166
ZStandard, 72, 76–77, 83–84
ZTrig, 79, 84
communicating with PC
  downloading TI Connect, 219–220
  installing and running
    TI Connect, 220
  TI-Graph Link cable and, 220
  transferring files, 220–221
Complex mode, 41–42
complex numbers
  conjugate of, finding, 43
  polar angle and modulus of,
    finding, 43–44
  real or imaginary part of,
    finding, 43
  setting mode for, 41–42
  using, 41–42
conj function, 43
Connected mode (Mode menu), 17
contrast, increasing or
    decreasing, 10

converting
  coordinates, 157–160
  between decimal and fraction,
    33, 158–159
  between degrees and DMS, 38–39
  degrees to radians, 37–38
  radians to degrees, 38
  between rectangular and polar
    form, 44
coordinates, converting, 157–160
CoordOn and CoordOff options
    (Format menu), 71, 86–87
copying
  files between calculator
    and PC, 221
  one matrix to another, 208–209
CPX submenu (Math menu)
  accessing, 42
  angle and abs functions, 43–44
  conj function, 43
  overview of, 32
  Rect and Polar functions, 44
cube function, 32, 33
cube-root function, 33
cursor, 11
curve fitting, 191

• *D* •

data
  analyzing, 197–199
  deleting and editing, 184–185
  entering, 183–184
  formula, using to enter, 187
  one-variable, plotting, 191–194
  saving and recalling data list,
    188–189
  sorting data list, 189–190
  tracing plots, 195–197
  two-variable, plotting, 195
  user-named data list, creating,
    185–187

dates, calculating number of days between two, 66
Dec function (Math menu), 33
decimals
   converting to fraction, 33, 158–159
   random, generating, 180–181
defining
   argument, 239
   matrices, 205–206
   parametric equation, 140
   solution bounds for Equation Solver, 47–48
definite integral, evaluating, 102–103
Degree setting (Mode menu)
   entering angle in radian measure in, 159
   overview of, 17
   trigonometric functions and, 236
degrees
   converting DMS to, 39
   converting radians to, 38
   converting to DMS, 38
   converting to radians, 37–38
   entering angle in while in Radian mode, 236
Del key, 13
ΔTbl value, 133, 150
deleting
   entry, 13
   Graph Database, 81
   matrix from memory, 209
   Picture, 114
   shading, 102
   statistical data, 184–185
derivatives, finding
   for functions, 100–101
   nDeriv function, 34
   for parametric equations, 155–156
   for polar equations, 175–176
determinant of matrix, finding, 214, 215
DiagnosticOn command, 200

differentiation, 34
displaying
   functions in table, 88–91
   matrices, 207
   parametric equations in table, 149–152
   polar equations in table, 169–172
   sequences in table, 132–135
distortion of graph, 76
DMS (degrees, minutes, seconds)
   converting degrees to, 38
   converting to degrees, 39
Dot mode (Mode menu), 17
Dotted Line style, 75, 77
downloading
   application from Texas Instruments Web site, 233
   applications, and battery low warning, 9–10
   TI Connect, 219–220
Draw menu. See also drawing on graph
   ClrDraw command, 113
   Pen option, 112
   Shade option, 109–111
   Text option, 111–112
Draw Points menu, 113
Draw Store menu, 113
drawing on graph
   circle, 107–108
   erasing, 112–113
   example of, 105
   freehand, 112
   function, 108–109
   horizontal or vertical line, 107
   inverse function, 109
   line segment, 106
   saving as Picture, 113–114
   sequence graph, 126
   tangent, 108
DuplicateName menu, 225, 226

• *E* •

e key, 22
editing. *See also* Stat List editor;
    Table Setup editor; Window
    editor; Y= editor
  definition of equation, 152
  definition of function in table, 91
  definition of sequence, 135
  entry, 13–14
  equation in Equation Solver, 47
  matrix, 207
  statistical data, 184–185
effective rate
  definition of, 53
  finding, given nominal rate,
    53–54
  finding nominal rate from, 54
embedding last answer in
    expression, 25
Engineering (Eng) mode, 16
Enter key, 12
entering
  angle in DMS measure, 39
  angle in radian measure, 159
  argument, 23–24
  argument for menu functions, 236
  arithmetic expression, 21–22
  equation in Equation Solver,
    46–47
  graphing function, 69–70
  identity matrix, 212, 213
  last entry, reusing, 25–26
  number, 22
  parametric equation, 139–141
  polar equation, 160–161
  scalar multiple of matrix, 212
  sequence, 117–120
  statistical data, 183–184
  statistical data using formula, 187

equation. *See also* Equation Solver;
    parametric equations;
    polar equations
  editing definition of, 152
  iterative, 17
  solving system of linear, 215–216
Equation Solver
  assigning value to variable, 47
  defining solution bounds, 47–48
  entering equation, 46–47
  finding multiple solutions, 49–50
  guessing solution, 48
  overview of, 45
  setting mode, 46
  solving equation, 49
erasing
  drawing, 112–113
  part of entry, 13
ERR: ARGUMENT error message,
    236, 239
ERR: BAD GUESS error message,
    48, 239–240
ERR: DATA TYPE error message,
    44, 213, 240
ERR: DIM MISMATCH error
    message, 212, 213, 240
ERR: DOMAIN error message, 240
ERR: INVALID DIM error message,
    71, 213, 214, 237, 240
ERR: INVALID error message,
    125, 240
ERR: NO SIGN CHNG error message,
    49, 241
ERR: SINGULAR MAT error
    message, 213, 241
ERR: SYNTAX error message,
    22, 235, 241
ERR: WINDOW RANGE error
    message, 237, 241

Error in Xmit error message, 223, 226
error messages, common, 239–241
errors
 common, 235–237
 round-off, 62
"escape" key equivalent, 12
evaluating
 arithmetic expression involving matrix, 211–214
 definite integral, 102–103
 expression, 22
 expression, order of operations and, 235–236
 function, 23–24
 function at specified value of *x,* 96
 parametric equation at specified value of T, 154–155
 permutation or combination, 179–180
 polar equation at specified value of θ, 174–175
 sequence at specified value of *n,* 137–138
exponential function key, 22–23
expression
 combining, 27–28
 entering and evaluating, 21–22
 inserting Math menu function into, 32–33
 inserting stored number into, 27
 recalling last, 14
 using previous answer in, 24–25
ExprOn and ExprOff options (Format menu), 72, 86–87

● *F* ●

Fibonacci sequence, 117
files, transferring
 between calculator and PC, 220–221
 between calculators, 224–226
 to several calculators, 226–227

financial goal, reaching, 63–64
Flash technology, 221, 224, 231
Float 0123456789 setting (Mode menu), 17
fMin and fMax functions, 34
fnInt function, 34–35
Format menu
 accessing, 71
 options, 71–72
 Parametric mode, 141–142
 Polar mode, 161–162
 Sequence mode, 121–122
formula, using to enter data, 187
fPart function, 36
Frac function (Math menu), 33
fractional part of value, finding, 36
fractions
 converting decimal to, 33, 158–159
 converting to decimal, 33
Full screen mode (Mode menu)
 overview of, 18
 viewing graph or table in, 136–137
Func (Function) mode (Mode menu)
 accessing, 69
 overview of, 17, 70
 panning in, 88
 Window editor, 72–73
function graphs
 setting different style for multiple, 73–75
 steps for, 70–73
 trigonometric, 79
functions. *See also* function graphs
 abs (Math menu), 35, 43
 angle (Math menu, CPX submenu), 43
 conj, 43
 cube, 32, 33
 cube-root, 33
 Dec (Math menu), 33
 displaying in table, 88–91
 drawing on graph, 108–109
 entering, 69–70
 fMin and fMax, 34

fnInt, 34–35
fPart, 36
Frac (Math menu), 33
gcd, 36
greatest-integer, 36
hyperbolic, 18–19
imag, 43
int, 36
iPart, 36
lcm, 36
Math menu, 32
max, 36
min, 36
nDeriv, 34
piecewise-defined, 77–79
Polar, 44
rational, 236
real, 43
Rect, 44
round, 35
shading area between, 109–111
slope (derivative), finding, 100–101
trigonometric, 18, 79, 236
value of, finding, 95–97
viewing on same screen as graph,
    79–80
xth root, 34
zeros of, finding, 97–98
future value of money, finding, 64–65

• G •

gcd function, 36
generating random numbers,
    180–181
GeoMaster application, 232
Graph Database, saving graph as,
    80–82, 126
graphs. *See also* function graphs;
    parametric graphs; polar
    graphs; sequence graphs
maximum or minimum point on,
    finding, 98–99
point of intersection of two,
    finding, 99–100
redrawing, 86, 129
saving and recalling, 80–82, 126
viewing on same screen as
    function, 79–80
viewing on same screen as table,
    92–93, 136–137
writing text on, 111–112
greatest common divisor, finding, 36
greatest-integer function, 36
GridOn and GridOff options
    (Format menu), 71
G-T mode (Mode menu)
overview of, 18
sequence and, 136
guessing solution, 239–240

• H •

highlighting item in menu, 15
histograms
constructing, 192–194
description of, 191, 192
tracing, 196, 197
Home screen
overview of, 13
returning to, 12
Horiz mode (Mode menu), 18
hyperbolic functions, 18–19

• I •

imag function, 43
Inequality Graphing application, 232
inserting
character, 14
Math menu function, 32–33
stored number into expression, 27
installing
application, 233
TI Connect, 220
int function, 36
integer part of value, finding, 36
integers, random, generating, 180

integration, 34–35
interest rate
  continuous compounding of, 59
  effective rate, finding, 53–54
  nominal rate, finding, 54
internal rate of return, finding, 55–56
inverse function, drawing on
    graph, 109
inverse function key, 23
inverse of matrix, finding, 213
inverse trigonometric function
    key, 22
iPart function, 36
iterative equation, 17

• K •

key strokes, 2
keyboard overview, 10
keying over existing character, 14
keys
  Alpha, 11
  Apps, 231
  Arrow, 10, 12
  Clear, 12, 13
  Del, 13
  e, 22
  Enter, 12
  exponential function, 22–23
  inverse function, 23
  inverse trigonometric function, 22
  negation, 22
  On, 10, 13, 163
  π, 22
  2nd (secondary function), 10, 11
  square function, 23
  square-root, 22–23
  subtraction, 22
  trigonometric function, 22
  X,T,Θ,n, 12

• L •

LabelOn and LabelOff options
    (Format menu), 72
last entry, reusing, 25–26
lcm function, 36
leasing versus borrowing, 55–56
least common multiple, finding, 36
left bound indicator, 99
letters, accessing, 11
line, horizontal or vertical, drawing
    on graph, 107
line segment, drawing on graph, 106
Line style, 75
Link Send menu, 224, 225, 227
linking calculators, 223
List Select menu, 224, 225
loans and mortgages
  balance, finding, 61
  principal and interest, finding, 60
  TVM (time-value-of-money) Solver,
    57–59
locking in Alpha mode, 11

• M •

Math menu. *See also* Equation
    Solver
  accessing, 14, 31
  CPX submenu, 32, 42–44
  functions, using, 32
  inserting function, 32–33
  MATH submenu, 31, 33–35
  NUM submenu, 31, 35–36
  overview of, 31
  PRB submenu, 32
  submenus of, 15
Math Probability menu
  evaluating permutation or
    combination, 179–180
  generating random numbers,
    180–181

MATH submenu (Math menu)
  accessing, 33
  cube and cube-root functions, 33
  Dec function, 33
  fMin and fMax functions, 34
  fnInt function, 34–35
  Frac function, 33
  nDeriv function, 34
  overview of, 31
  *x*th root function, 34
matrices
  augmenting two, 208
  copying one to another, 208–209
  defining, 205–206
  deleting from memory, 209
  description of, 205
  displaying, 207
  editing, 207
  finding determinant of, 214–215
  solving system of linear equations,
    215–216
  using in arithmetic expression,
    211–214
Matrix editor menu, accessing, 205
Matrix Math menu
  augment command, 207
  Transpose command, 214
max function, 36
maximum point on graph, finding,
  98–99
maximum value
  finding in list of numbers, 36
  finding location of, 34
Memory Management menu, 185
menu function, entering argument
    for, 236
menus. *See also* Format menu;
    Math menu; Mode menu
  accessing, 14–15
  Angle, 37–40, 157

Calculate (CALC), 95–102, 155–156
Draw, 109–113
Draw Points, 113
Draw Store, 113
DuplicateName, 225, 226
Link Send, 224, 225, 227
List Select, 224, 225
Math Probability, 179–181
Matrix editor, 205
Matrix Math, 207, 214
Memory Management, 185
scrolling, 15
selecting item from, 15
Stat Calculate, 198, 200
Stat Plots, 192
Test, 14
Zoom, 127
Zoom Memory, 129
min function, 36
minimum point on graph, finding,
  98–99
minimum value
  finding in list of numbers, 36
  finding location of, 34
mini-program, writing, 28
Mode menu
  a + b*i* mode, 41–42
  Complex mode, 41–42
  Degree mode, 159
  Full Screen mode, 136–137
  Function mode, 69, 70, 72–73, 88
  G-T mode, 136
  overview of, 16–18
  Parametric mode, 140,
    142–143, 145
  Polar mode, 160, 161–163, 169
  Radian mode, 159, 236
  selecting item from, 15
  Sequential mode, 118, 122–123,
    127–129, 131

modified box plot
  description of, 194
  tracing, 196
modulus of complex number,
      finding, 43–44
multiplying matrices, 213

• *N* •

naming data list, 185–187
nCr (combination), 179–180
nDeriv function, 34
negating
  matrix, 212
  number, 235
negation key, 22
negative cash flow, distinguishing
      from positive, 56
*n*Max, setting for sequence, 123
*n*Min, setting for sequence,
      118, 123
nominal rate
  definition of, 53
  finding effective rate from, 53–54
  finding, given effective rate, 54
Normal mode, 16
Normal, Sci, or Eng setting
      (Mode menu), 16
NoteFolio application, 232
nPr (permutation), 179–180
NUM submenu (Math menu)
  functions, 35–36
  overview of, 31
numbers. *See also* complex
      numbers
  entering, 22
  negating, 235
  random, generating, 180–181
  rounding, 35

• *O* •

On key, 10, 13, 163
one-variable data analysis, 197–198
one-variable data, plotting, 191–194
order of operations, 23–24, 235–236
Organizer application, 232
outlier, modified box plot and, 194
overriding mode of angle, 39–40

• *P* •

panning
  in Function mode, 88
  in Parametric mode, 145
  in Polar mode, 169
  in Sequence mode, 131
parametric equations. *See also*
      parametric graphs
  definition of, 148
  derivative, finding, 155–156
  displaying in table, 149–152
  entering, 139–141
  evaluating, 154–155
  graphing, 141–144
  independent variable T, 148–149
  overview of, 139
  setting different styles for multiple
      graphs, 144–145
  values of *x* and *y,* 149
  Zoom commands and, 146–147
parametric graphs
  saving, 148
  tracing, 148–149
  viewing on same screen as table,
      153–154
Parametric (Par) mode
      (Mode menu)
  accessing, 140
  overview of, 17

panning in, 145
Window editor and, 142–143
parentheses, use of, 23–24, 235–236
pasting
function name in another
function, 70
name of matrix into
expression, 212
parametric equation name in
another equation, 140
polar equation name in another
equation, 160
value from TVM Solver into other
expression, 59–60
Path style, 74, 75, 145
PC, communicating with
downloading TI Connect, 219–220
installing and running
TI Connect, 220
TI-Graph Link cable and, 220
transferring files, 220–221
permutation (nPr), 179–180
π key, 22
picture of graph, saving, 80
piecewise-defined function, 77–79
PlotStart, setting for sequence,
123–124
PlotStep, setting for sequence, 124
plotting
one-variable data, 191–194
two-variable data, 195
point of intersection, finding, 99–100
point, representing in polar
coordinate system, 157
Pol (Polar) setting (Mode menu)
accessing, 160
Format menu and, 161–162
overview of, 17
panning in, 169
Window editor and, 162–163

polar angle of complex number,
finding, 43–44
polar coordinates
converting from rectangular
coordinates to, 157–158
converting to rectangular
coordinates, 159–160
overview of, 157
polar equations. *See also* polar
graphs
converting coordinates to and
from, 157–160
derivatives, finding, 175–176
displaying in table, 169–172
entering, 160–161
evaluating at specified value of θ,
174–175
graphing, 161–164
independent variable θ, 168
values *x* and *y,* 168
polar form
converting to rectangular form, 44
entering number in, 42
Polar function, 44
polar graphs
saving, 167
setting different styles for multiple,
164–165
tracing, 167–169
viewing on same screen as
table, 172
PolarGC option (Format menu)
function and, 71, 87
sequence and, 130
positive cash flow, distinguishing
from negative, 56
PRB submenu (Math menu), 32
present value of money, finding,
65–66
pressing keys, 10

previous answer, using in
    expression, 24–25
principal of loan, finding, 61
Probability Simulation
    application, 232

• *Q* •

Quit command, 13

• *R* •

r (distance), 157
radian measure, converting to
    fractional multiple of π, 158–159
Radian setting (Mode menu)
  entering angle in degree
    measure in, 159
  entering angle in degrees
    while in, 236
  overview of, 17, 18
radians
  converting degrees to, 37–38
  converting to degrees, 38
raising matrix to positive integral
    power, 213–214
random numbers, generating,
    180–181
rational function, 236
real function, 43
Real mode (Mode menu), 18
recalling
  data list, 188–189
  last expression or command, 14
  Picture, 114
  saved Graph Database, 80–82
re^θi mode (Mode menu), 18, 41–42
Receiving mode, 224
Rect function, 44

rectangular coordinates
  converting from polar coordinates
    to, 159–160
  converting to polar coordinates
    from, 157–158
rectangular form
  a + bi mode and, 18, 41–42
  converting to polar form, 44
  entering number in, 42
RectGC option (Format menu)
  function and, 71
  sequence and, 121–122
redrawing graph
  ZBox command and, 86
  ZPrevious command and, 129
reduced row-echelon form, 215–216
regression modeling, 199–201
renaming file being transferred,
    225–226
returning to Home screen, 12
reusing
  command, 54
  last entry, 25–26
right bound indicator, 99
round function, 35
round-off error, 62
rref command, 216
running TI Connect, 220

• *S* •

saving
  data list, 188–189
  function graph, 80–82
  graph as Graph Database,
    80–82, 126
  graph as Picture, 113–114
  parametric graph, 148
  polar graph, 167
  sequence graph, 126

scatter plots
  description of, 195
  tracing, 196, 197
Scientific (Sci) mode, 16
scrolling menu, 15
2nd (secondary function) key, 10, 11
selecting item from menu, 15
Seq (Sequential) mode
    (Mode menu)
  accessing, 118
  overview of, 17
  panning in, 131
  Window editor and, 122–123
  Zoom commands and, 127–129
sequence graphs
  drawing on, 126
  saving, 126
  setting different styles for
    multiple, 125–126
  viewing on same screen as table,
    136–137
sequences. *See also* sequence
    graphs
  definition of, 129
  description of, 117
  displaying in table, 132–135
  editing definition of, 135
  entering, 117–120
  evaluating at specified value of *n*,
    137–138
  graphing, 121–125
  independent variable *n*, 129–130
  tracing, 129–132
SetUpEditor command, 183, 188
shading
  area between functions, 109–111
  erasing, 102
  patterns of, 74, 75
ΣInt command, 60
ΣPrn command, 60
Simultaneous (Simul) mode
    (Mode menu), 17

size of viewing window, 76–77, 84
slope of curve, finding, 100–101
Solver
  assigning value to variable, 47
  defining solution bounds, 47–48
  entering equation, 46–47
  ERR: NO SIGN CHNG error
    message, 241
  finding multiple solutions, 49–50
  guessing solution, 48
  overview of, 45
  setting mode, 46
  solving equation, 49
solving system of linear equations,
    215–216
sorting data list, 189–190
square function key, 23
square-root key, 22–23
Standard viewing window, 72
starting expression with last answer,
    24–25
Stat Calculate menu
  one-variable data analysis, 198
  regression model, 200
Stat List editor
  deleting and editing data, 184–185
  entering data, 183–184
  formula, using to enter data, 187
  saving and recalling data list,
    188–189
  user-named data list, creating,
    185–187
Stat Plots
  common errors involving, 237
  menu, accessing, 192
  turning off highlighted in
    Y= editor, 71
statistical data
  analyzing, 197–199
  deleting and editing, 184–185

statistical data *(continued)*
   entering, 183–184
   formula, using to enter data, 187
   one-variable data, plotting, 191–194
   saving and recalling data list,
      188–189
   sorting data list, 189–190
   tracing plots, 195–197
   two-variable data, plotting, 195
   user-named data list, creating,
      185–187
storing variable, 26–27
StudyCards application, 232
style of graph, setting
   for function, 73–75
   for parametric equation, 144–145
   for polar equation, 164–165
   for sequence, 125–126
subtracting matrices, 212–213
subtraction key, 22
system of linear equations, solving,
      215–216

• *T* •

t (parameter), 139
Table Setup editor
   function and, 89
   parametric equation and, 150
   sequence and, 132
tables. *See also* automatically
      generated table; user-generated
      table
   clearing, 91–92
   displaying functions in, 88–91
   displaying parametric equations
      in, 149–152
   displaying polar equations in,
      169–172
   displaying sequences in, 132–135
   ERROR display, 135, 152

viewing on same screen as
      function, 92–93
viewing on same screen as
      parametric graph, 153–154
viewing on same screen as polar
      graph, 172–173
viewing on same screen as
      sequence, 136–137
tangent, drawing on graph, 108
TblStart value, 132, 134, 150
terminating graphing process,
      13, 163
Test menu, accessing, 14
Texas Instruments Web site
   downloading applications from,
      232–233
   online store, 219
text, writing on graph, 111–112
θ (angular direction), 157
θmax variable (Polar mode), 162
θmin variable
      (Polar mode), 162
θstep variable
      (Polar mode), 162–163
Thick Line style, 75
TI Connect
   downloading, 219–220
   installing and running, 220
   transferring files, 220–221
   upgrading operating system
      sand, 221
TI-Graph Link cable, 219, 220
Time format, 121, 130, 138
Tmax variable (Window editor),
      142, 143
Tmin variable (Window editor),
      142, 143
tracing
   graph, 86–88
   parametric graph, 148–149
   polar graph, 167–169

sequence, 129–132
statistical data plot, 195–197
transferring files
  between calculator and PC,
    220–221
  between calculators, 224–226
  to several calculators, 226–227
Transpose command, 214
transposing matrix, 214, 215
trigonometric function
  Degree mode and, 236
  graphing, 79
  Radian mode and, 18
trigonometric function key, 22
Tstep variable (Window editor),
    142–143
turning on and off, 10
TVM (time-value-of-money) Solver
  balance of loan, finding, 61
  financial goal, reaching, 63–64
  future value of money, finding,
    64–65
  overview of, 57–59
  pasting value from into other
    expression, 59–60
  present value of money, finding,
    65–66
  principal and interest, finding, 60
two-variable data
  plotting, 195
  regression model for, 200–201
two-variable data analysis, 197, 199

• *U* •

undoing Zoom command, 87
unit-to-unit link cable, 223
u(*n*Min), setting for sequence,
    119–120
upgrading operating system, 221

user-generated table
  clearing, 91, 135, 152
  creating, 133
  parametric equation and, 151, 152
  polar equations and, 171–172
  TblStart value and, 134
  terminating process of, 134–135
user-named data list, creating,
    185–187
uv format, 122, 131, 138
uw format, 122, 131, 138

• *V* •

value command, 96–97, 138
values
  absolute, finding, 35
  assigning to variable, 47
  finding in list of numbers, 36
  finding location of, 34
  fractional part of, finding, 36
  of function, finding, 95–97
  future, of money, finding, 64–65
  integer part of, finding, 36
  present, of money, finding, 65–66
variables
  assigning value to in Equation
    Solver, 47
  entering, 12
  one-variable data analysis,
    197–198
  one-variable data, plotting,
    191–194
  regression modeling, 199–201
  storing, 26–27
  two-variable data analysis,
    197, 199
  two-variable data, plotting, 195
vertical asymptotes, eliminating, 77

viewing. *See also* viewing window
  function and graph on same
      screen, 79–80
  parametric graph and table on
      same screen, 153–154
  polar graph and table on same
      screen, 172–173
  readjusting after using
      ZoomFit, 144
  table and graph on same screen,
      92–93, 136–137
viewing window
  adjusting to center on cursor
      location, 88
  histogram and, 193–194
  setting, 76–77
  Standard, 72
  ZBox command and, 127–128
  Zoom commands and, 83–86
vw format, 122, 131, 138

• W •

Web format, 122, 130, 131, 138
web plot, 122
Web site, Texas Instruments,
      219, 232–233
Window editor
  ERR: WINDOW RANGE error
      message and, 237
  Function mode, 72–73
  histogram and, 193–194
  Parametric mode, 142–143
  Polar mode, 162–163
  Sequence mode, 122–123
writing
  mini-program, 28
  text on graph, 111–112

• X •

Xmin and Xmax variables
      (Window editor)
  overview of, 72, 87, 88
  readjusting after using ZoomFit
      command, 144
  sequence, setting for, 124
Xres variable (Window editor), 73
Xscl variable (Window editor)
  overview of, 73, 124
  readjusting after using ZoomFit
      command, 144
xth root function, 34
X,T,Θ,n key, 12
xy-line plot, 195, 196

• Y •

Y= editor
  accessing, 140
  delaying graphing of function in, 75
  entering functions into, 70
  overview of, 32
  turning off highlighted Stat Plot in,
      71, 121, 141
Ymin and Ymax variables
      (Window editor)
  overview of, 73, 87, 88
  readjusting after using ZoomFit
      command, 144
  sequence, setting for, 125
Yscl variable (Window editor)
  overview of, 73, 125
  readjusting after using ZoomFit
      command, 144

• Z •

ZBox command
  functions and, 85–86
  parametric equations and, 146–147
  polar equations and, 166
  sequences and, 127–128
ZDecimal command, 84
zeros of function, finding, 97–98
ZInteger command, 85
Zoom commands. *See also specific commands*
  overview of, 69, 83
  parametric equations and, 146–147
  polar equations and, 165–167
  Sequence mode and, 127–129
  undoing, 87
  using, 83–85
Zoom In command
  functions and, 83–84, 85
  parametric equations and, 147
  polar equations and, 167
  sequences and, 128–129
Zoom Memory menu, 129
Zoom menu, accessing, 127

Zoom Out command
  functions and, 83–84, 85
  parametric equations and, 147
  polar equations and, 167
  sequences and, 128–129
ZoomFit command
  functions and, 73, 84
  parametric equations and, 146
  Parametric mode, 143
  Polar mode, 163, 165
  readjusting viewing window after using, 144
  sequences and, 124, 127
ZoomStat command
  functions and, 84
  histograms and, 193
  scatter or *xy*-line plots, 195
ZPrevious command, 129, 148
ZSquare command
  functions and, 76, 84–85
  parametric equations and, 146
  polar equations and, 165–166
ZStandard command, 72, 76–77, 83–84
ZTrig command, 79, 84

Notes

Notes

....

Notes

....

_____

_____

_____

_____

_____

_____

_____

_____

_____

_____

_____

_____

_____

_____

_____

_____

_____

_____

_____

FOR DUMMIES®

The easy way to get more done and have more fun

PERSONAL FINANCE

0-7645-5231-7

0-7645-2431-3

0-7645-5331-3

Also available:

Estate Planning For Dummies
(0-7645-5501-4)

401(k)s For Dummies
(0-7645-5468-9)

Frugal Living For Dummies
(0-7645-5403-4)

Microsoft Money "X" For Dummies
(0-7645-1689-2)

Mutual Funds For Dummies
(0-7645-5329-1)

Personal Bankruptcy For Dummies
(0-7645-5498-0)

Quicken "X" For Dummies
(0-7645-1666-3)

Stock Investing For Dummies
(0-7645-5411-5)

Taxes For Dummies 2003
(0-7645-5475-1)

BUSINESS & CAREERS

0-7645-5314-3

0-7645-5307-0

0-7645-5471-9

Also available:

Business Plans Kit For Dummies
(0-7645-5365-8)

Consulting For Dummies
(0-7645-5034-9)

Cool Careers For Dummies
(0-7645-5345-3)

Human Resources Kit For Dummies
(0-7645-5131-0)

Managing For Dummies
(1-5688-4858-7)

QuickBooks All-in-One Desk Reference For Dummies
(0-7645-1963-8)

Selling For Dummies
(0-7645-5363-1)

Small Business Kit For Dummies
(0-7645-5093-4)

Starting an eBay Business For Dummies
(0-7645-1547-0)

HEALTH, SPORTS & FITNESS

0-7645-5167-1

0-7645-5146-9

0-7645-5154-X

Also available:

Controlling Cholesterol For Dummies
(0-7645-5440-9)

Dieting For Dummies
(0-7645-5126-4)

High Blood Pressure For Dummies
(0-7645-5424-7)

Martial Arts For Dummies
(0-7645-5358-5)

Menopause For Dummies
(0-7645-5458-1)

Nutrition For Dummies
(0-7645-5180-9)

Power Yoga For Dummies
(0-7645-5342-9)

Thyroid For Dummies
(0-7645-5385-2)

Weight Training For Dummies
(0-7645-5168-X)

Yoga For Dummies
(0-7645-5117-5)

Available wherever books are sold.
Go to www.dummies.com or call 1-877-762-2974 to order direct.

FOR DUMMIES®

A world of resources to help you grow

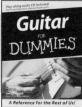

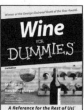

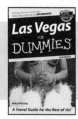

FOR DUMMIES®

Helping you expand your horizons and realize your potential

INTERNET

0-7645-0894-6

0-7645-1659-0

0-7645-1642-6

Also available:

America Online 7.0 For Dummies
(0-7645-1624-8)

Genealogy Online For Dummies
(0-7645-0807-5)

The Internet All-in-One Desk Reference For Dummies
(0-7645-1659-0)

Internet Explorer 6 For Dummies
(0-7645-1344-3)

The Internet For Dummies Quick Reference
(0-7645-1645-0)

Internet Privacy For Dummies
(0-7645-0846-6)

Researching Online For Dummies
(0-7645-0546-7)

Starting an Online Business For Dummies
(0-7645-1655-8)

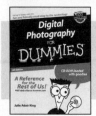

0-7645-1664-7

0-7645-1675-2

0-7645-0806-7

Also available:

CD and DVD Recording For Dummies
(0-7645-1627-2)

Digital Photography All-in-One Desk Reference For Dummies
(0-7645-1800-3)

Digital Photography For Dummies Quick Reference
(0-7645-0750-8)

Home Recording for Musicians For Dummies
(0-7645-1634-5)

MP3 For Dummies
(0-7645-0858-X)

Paint Shop Pro "X" For Dummies
(0-7645-2440-2)

Photo Retouching & Restoration For Dummies
(0-7645-1662-0)

Scanners For Dummies
(0-7645-0783-4)

GRAPHICS

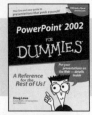

0-7645-0817-2

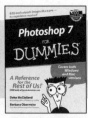

0-7645-1651-5

0-7645-0895-4

Also available:

Adobe Acrobat 5 PDF For Dummies
(0-7645-1652-3)

Fireworks 4 For Dummies
(0-7645-0804-0)

Illustrator 10 For Dummies
(0-7645-3636-2)

QuarkXPress 5 For Dummies
(0-7645-0643-9)

Visio 2000 For Dummies
(0-7645-0635-8)

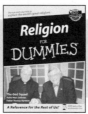

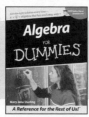

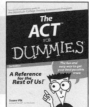

FOR DUMMIES

We take the mystery out of complicated subjects